ENERGY SECTOR STANDARD
OF THE PEOPLE'S REPUBLIC OF CHINA

中华人民共和国能源行业标准

Code for Seismic Design of Hydraulic Structures
of Hydropower Project

水电工程水工建筑物抗震设计规范

NB 35047-2015
Replace DL 5073-2000

Chief Development Department: China Renewable Energy Engineering Institute
Approval Department: National Energy Administration of the People's Republic of China
Implementation Date: September 1, 2015

China Water & Power Press

Beijing 2023

All rights reserved. No part of this publication may be reproduced, stored in a retrieval system, or transmitted in any form or by any means—electronic, mechanical, photocopying, recording or otherwise, without prior written permission of the publisher.

图书在版编目（CIP）数据

水电工程水工建筑物抗震设计规范 : NB 35047-2015 = Code for Seismic Design of Hydraulic Structures of Hydropower Project（NB 35047-2015）: 英文 / 国家能源局发布. -- 北京 : 中国水利水电出版社, 2024. 8. ISBN 978-7-5226-2717-5

Ⅰ．TV6-65

中国国家版本馆CIP数据核字第2024Z5L887号

ENERGY SECTOR STANDARD
OF THE PEOPLE'S REPUBLIC OF CHINA
中华人民共和国能源行业标准

Code for Seismic Design of Hydraulic Structures
of Hydropower Project
水电工程水工建筑物抗震设计规范

NB 35047-2015

Replace DL 5073-2000

（英文版）

Issued by National Energy Administration of the People's Republic of China
国家能源局　发布
Translation organized by China Renewable Energy Engineering Institute
水电水利规划设计总院　组织翻译
Published by China Water & Power Press
中国水利水电出版社　出版发行
　　Tel: (+ 86 10) 68545888　68545874
　　sales@mwr.gov.cn
　　Account name: China Water & Power Press
　　Address: No.1, Yuyuantan Nanlu, Haidian District, Beijing 100038, China
　　http: //www.waterpub.com.cn
中国水利水电出版社微机排版中心　排版
北京中献拓方科技发展有限公司　印刷
184mm×260mm　16开本　5印张　158千字
2024年8月第1版　2024年8月第1次印刷

Price（定价）：￥770.00

Introduction

The English version is one of China's energy sector standard series in English. Its translation was organized by China Renewable Energy Engineering Institute authorized by National Energy Administration of the People's Republic of China in compliance with relevant procedures and stipulations. This English version was issued by National Energy Administration of the People's Republic of China in Announcement [2021] No.5 dated November 16, 2021.

This version was translated from the Chinese Standard NB 35047-2015, *Code for Seismic Design of Hydraulic Structures of Hydropower Project*, published by China Electric Power Press. The copyright is reserved by National Energy Administration of the People's Republic of China. In the event of any discrepancy in the implementation, the Chinese version shall prevail.

Many thanks go to the staff from the relevant standard development organizations and those who have provided generous assistance in the translation and review process.

For further improvement of the English version, any comments and suggestions are welcome and should be addressed to:

China Renewable Energy Engineering Institute
No. 2 Beixiaojie, Liupukang, Xicheng District, Beijing 100120, China
Website: www.creei.cn

Translating organization:

China Institute of Water Resources and Hydropower Research
China Renewable Energy Engineering Institute

Translating staff:

CHEN Houqun	LI Deyu	HU Xiao	LIU Xiaosheng
WANG Haibo	ZHAO Jianming	ZHANG Yanhong	ZHANG Boyan
TU Jin	ZHANG Cuiran	OUYANG Jinhui	GUO Shengshan
YANG Zhengquan	YANG Yusheng	LIU Rongli	WU Mingxin
ZHONG Hong	LIANG Hui		

Review panel members:

JIN Feng	Tsinghua University
GUO Jie	POWERCHINA Beijing Engineering Corporation Limited
YAN Wenjun	Army Academy of Armored Forces, PLA

LIU Qing	POWERCHINA Northwest Engineering Corporation Limited
JIA Haibo	POWERCHINA Kunming Engineering Corporation Limited
LI Jinrong	POWERCHINA Huadong Engineering Corporation Limited
LI Zhaojin	POWERCHINA Zhongnan Engineering Corporation Limited
CHENG Jing	Hohai University
ZHANG Ming	Tsinghua University
LIU Xiaofen	POWERCHINA Zhongnan Engineering Corporation Limited
QIAO Peng	POWERCHINA Northwest Engineering Corporation Limited

National Energy Administration of the People's Republic of China

翻译出版说明

本译本为国家能源局委托水电水利规划设计总院按照有关程序和规定，统一组织翻译的能源行业标准英文版系列译本之一。2021年11月16日，国家能源局以2021年第5号公告予以公布。

本译本是根据中国电力出版社出版的《水电工程水工建筑物抗震设计规范》NB 35047—2015 翻译的，著作权归国家能源局所有。在使用过程中，如出现异议，以中文版为准。

本译本在翻译和审核过程中，本标准编制单位及编制组有关成员给予了积极协助。

为不断提高本译本的质量，欢迎使用者提出意见和建议，并反馈给水电水利规划设计总院。

地址：北京市西城区六铺炕北小街2号
邮编：100120
网址：www.creei.cn

本译本翻译单位：中国水利水电科学研究院
　　　　　　　　　水电水利规划设计总院

本译本翻译人员：陈厚群　李德玉　胡　晓　刘小生
　　　　　　　　　王海波　赵剑明　张艳红　张伯艳
　　　　　　　　　涂　劲　张翠然　欧阳金惠　郭胜山
　　　　　　　　　杨正权　杨玉生　刘荣丽　武明鑫
　　　　　　　　　钟　红　梁　辉

本译本审核人员：

　金　峰　清华大学

　郭　洁　中国电建集团北京勘测设计研究院有限公司

　闫文军　中国人民解放军陆军装甲兵学院

　柳　青　中国电建集团西北勘测设计研究院有限公司

　贾海波　中国电建集团昆明勘测设计研究院有限公司

　李金荣　中国电建集团华东勘测设计研究院有限公司

　李兆进　中国电建集团中南勘测设计研究院有限公司

程　井　河海大学
张　明　清华大学
刘小芬　中国电建集团中南勘测设计研究院有限公司
乔　鹏　中国电建集团西北勘测设计研究院有限公司

国家能源局

Announcement of National Energy Administration of the People's Republic of China [2015] No. 3

According to the requirements of Document GNJKJ [2009] No. 52 "Notice on Releasing the Energy Sector Standardization Administration Regulations (*tentative*) and detailed implementation rules issued by National Energy Administration of the People's Republic of China", 203 sector standards such as *Carbon Steel and Low Alloy Steel for Pressurized Water Reactor Nuclear Power Plants—Part 31: 15Mn Forgings for Containment Vessel*, including 106 energy standards (NB) and 97 electric power standards (DL), are issued by National Energy Administration of the People's Republic of China after due review and approval.

Attachment: Directory of Sector Standards

National Energy Administration of the People's Republic of China

April 2, 2015

Attachment:

Directory of Sector Standards

Serial Number	Standard No.	Title	Replaced standard No.	Adopted international standard No.	Approval date	Implementation date
...						
68	NB 35047-2015	Code for Seismic Design of Hydraulic Structures of Hydropower Project	DL 5073-2000		2015-04-02	2015-09-01
...						

Foreword

According to the requirements of Document FGBGY [2008] No. 1242 2008-33 issued by General Office of the National Development and Reform Commission, "Notice on Releasing the Development and Revision Plan of the Sector Standards in 2008", and after extensive investigation and research, summarization of practical experience, and wide solicitation of opinions, the drafting group has prepared this code.

The main technical contents of this code include: seismic design criteria, ground motion parameter determination, seismic actions and calculation, seismic measures, etc., for dam and other main hydraulic structures.

The main technical contents revised in this code are as follows:

— Modifying the title of the code to "Code for Seismic Design of Hydraulic Structures of Hydropower Project".

— Transforming the safety factor design method to the limit state design method expressed by partial factors, according to the requirements for reliability design criteria in GB 50153-2008, *Unified Standard for Reliability Design of Engineering Structures* and GB 50199-2013, *Unified Standard for Reliability Design of Hydraulic Engineering Structures*, introducing the structural factor γ_d, which represents the non-random uncertainties, into the limit state design expression, and giving the corresponding values of seismic structural factor γ_d for various hydraulic structures.

— Revising the parameters of standard design response spectrum on bedrock for seismic motion of ordinary projects.

— Adding the following requirements: for hydraulic structures with seismic fortification Class A of which the design seismic parameters shall be provided by the special site seismic hazard analysis, special study under the maximum credible earthquake (MCE) shall be carried out on disaster prevention of the uncontrolled release of reservoir in addition to the seismic design under design peak ground acceleration (PGA). The MCE of the site shall be determined by the deterministic method or the probabilistic method with an exceeding probability of 1 % within 100 years of the reference period.

— Specifying considerations of dynamic analysis for concrete gravity dams and arch dams requiring seismic analysis under MCE. The following factors should be taken into account in the dynamic analysis model: dynamic interaction of structure-foundation-reservoir system, the mass of foundation, rock type and geological conditions of near-field foundation, the far-field radiation damping, non-uniform ground motion, material nonlinearity of dam concrete and near-field foundation, opening and sliding of arch dam transverse joints during earthquakes.

— Specifying the requirements that the finite element method (FEM) shall be used in dynamic analysis and safety evaluation for embankment dam body and foundation under certain conditions.

— Revising the characteristic values of dynamic strength and elastic modulus of dam concrete.

— Adding the seismic design of aqueduct, shiplift and slope.

National Energy Administration of the People's Republic of China is in charge of the administration of this code. China Renewable Energy Engineering Institute has proposed the code and is responsible for its routine management. Energy Sector Standardization Technical Committee on Hydropower Investigation and Design is responsible for the explanation of specific technical content. Comments and suggestions in the implementation of this code should be addressed to:

China Renewable Energy Engineering Institute
No. 2 Beixiaojie, Liupukang, Xicheng District, Beijing 100120, China

Chief development organizations:

China Renewable Energy Engineering Institute

China Institute of Water Resources and Hydropower Research

Chief drafting staff:

CHEN Houqun	DANG Lincai	LI Deyu	DU Xiaokai
HU Xiao	YAN Yongpu	LIU Xiaosheng	WANG Haibo
ZHAO Jianming	LI Guangshun	ZHANG Yanhong	ZHANG Boyan
WANG Zhongning	TU Jin	LI Min	ZHANG Cuiran
OUYANG Jinhui	MA Huaifa	GUO Shengshan	YANG Zhengquan

Review panel members:

WANG Minhao	ZHOU Jianping	LI Sheng	LI Shisheng
SUN Baoping	HU Bin	ZHANG Chuhan	TONG Xianwu
LYU Mingzhi	DENG Yiguo	WANG Renkun	WU Guanye
FAN Fuping	HONG Yongwen	YAO Shuanxi	XIAO Feng
AI Yongping	YAN Jun	LIN Peng	

Contents

1	**General**	**1**
2	**Terms and Symbols**	**3**
2.1	Terms	3
2.2	Symbols	5
3	**Basic Requirements**	**8**
4	**Site, Foundation and Slope**	**11**
4.1	Site	11
4.2	Foundation	13
4.3	Slope	14
5	**Seismic Action and Seismic Calculation**	**16**
5.1	Ground Motion Components and Combination	16
5.2	Classification of Seismic Actions	16
5.3	Design Response Spectra	17
5.4	Combination of Seismic Action and Other Actions	18
5.5	Structural Modeling and Calculation Method	18
5.6	Dynamic Properties of Concrete and Foundation Rock Mass	21
5.7	Seismic Design for Ultimate Limit States with Partial Factors	21
5.8	Seismic Calculation for Appurtenant Structure	23
5.9	Seismic Earth Pressure	23
6	**Embankment Dam**	**25**
6.1	Seismic Calculation	25
6.2	Seismic Measures	27
7	**Gravity Dam**	**30**
7.1	Seismic Calculation	30
7.2	Seismic Measures	33
8	**Arch Dam**	**35**
8.1	Seismic Calculation	35
8.2	Seismic Measures	38
9	**Sluice**	**39**
9.1	Seismic Calculation	39
9.2	Seismic Measures	41
10	**Underground Hydraulic Structure**	**43**
10.1	Seismic Calculation	43
10.2	Seismic Measures	45

11	**Intake Tower**	**46**
11.1	Seismic Calculation	46
11.2	Seismic Measures	50
12	**Penstock and Surface Powerhouse**	**52**
12.1	Penstocks	52
12.2	Surface Powerhouses	53
13	**Aqueduct**	**54**
13.1	Seismic Calculation	54
13.2	Seismic Measures	55
14	**Shiplift**	**56**
14.1	Seismic Calculation	56
14.2	Seismic Measures	56
Appendix A	Seismic Stability Calculation of Embankment Dams with Quasi-Static Method	58
Appendix B	Calculation of Hydrodynamic Pressure in Aqueduct	61
Explanation of Wording in This Code		**65**
List of Quoted Standards		**66**

1 General

1.0.1 This code is formulated in accordance with the *Law of the People's Republic of China on Protecting Against and Mitigating Earthquake Disasters*, and with a view to carrying out the policy of prevention first, to mitigate earthquake damage and prevent secondary disasters through seismic design of hydraulic structures.

1.0.2 This code is applicable to seismic design of Grade 1, 2 and 3 hydraulic structures with the design intensity of Ⅵ to Ⅸ, such as the roller-compacted embankment dam, concrete gravity dam, concrete arch dam, sluices, underground hydraulic structure, intake tower, penstock and surface powerhouse of hydropower station, aqueduct, shiplift, etc.

For hydraulic structures with design intensity of Ⅵ, seismic calculation need not be conducted, but appropriate seismic measures shall be taken in accordance with this code.

For hydraulic structures with design intensity over Ⅸ, water-retaining structures higher than 200 m or with particular problems, special study and demonstration shall be carried out on their seismic safety.

1.0.3 The design PGA on the project site and corresponding design intensity shall be determined as follows:

    1   For ordinary projects, the design PGA and intensity are determined in accordance with GB 18306, *Seismic Ground Motion Parameter Zonation Map of China*.

    2   **For large-scale (Rank 1) projects with a dam height over 200 m or reservoir storage capacity over 10 billion m³ in the regions with a basic intensity of Ⅵ or above, and large-scale (Rank 1) projects with a dam height over 150 m in the regions with a basic intensity of Ⅶ or above, the design PGA and intensity are determined based on site-specific seismic safety evaluation.**

    3   For Grade 1 and 2 dams 100 m to 150 m high and with complicated geological conditions in the region with a basic intensity of Ⅶ or above, the design PGA and intensity should be determined based on site-specific seismic safety evaluation.

1.0.4 The hydraulic structure designed as per this code shall be able to resist the seismic action of the design intensity, and can operate normally after repair in case of local damages.

1.0.5 In addition to this code, the seismic design of hydraulic structures shall comply with other current relevant standards of China.

2 Terms and Symbols

2.1 Terms

2.1.1 seismic design

special design of engineering structures in intensive seismic regions, generally including seismic calculation and seismic measures

2.1.2 basic intensity

seismic intensity of an ordinary site with an exceeding probability of 10 % for a 50-year period, determined by the site PGA specified in GB 18306 and the corresponding seismic intensity specified in its appendix

2.1.3 design intensity

seismic intensity for engineering fortification determined on the basis of the basic intensity

2.1.4 reservoir earthquake

earthquake related to reservoir impounding, which generally occurs less than 10 km away from the reservoir banks

2.1.5 maximum credible earthquake (MCE)

earthquake with potential maximum ground motion assessed based on the regional geological and seismological conditions around project site

2.1.6 scenario earthquake

earthquake with definite magnitude and epicenter distance, which is selected under the principle of maximum probability of occurrence along the main fault among the potential seismic sources that can generate design PGA on project site

2.1.7 seismic ground motion

ground motion induced by earthquake

2.1.8 seismic actions

dynamic actions of seismic ground motion on structures

2.1.9 hanging wall effect

phenomenon that seismic ground motion of hanging wall above the inclined seismogenic fault is larger than that of footwall

2.1.10 peak ground acceleration (PGA)

maximum absolute value of ground mass point motion acceleration during earthquake

2.1.11 design earthquake

ground motion for seismic fortification corresponding to design intensity, of which the parameters include PGA, response spectrum, duration, and acceleration time history

2.1.12 design seismic acceleration

peak acceleration of ground motion with probability level specified by site-specific seismic safety evaluation on project site, or generally corresponding to the design intensity

2.1.13 seismic effect

dynamic effect such as structure internal force, deformation, sliding, and cracking etc caused by seismic action

2.1.14 seismic liquefaction

phenomenon in which saturated cohesionless soil or less cohesive soil tends to be denser, the pore water pressure increases, and the effective stress approaches zero induced by the seismic ground motion

2.1.15 design response spectrum

response spectrum adopted in seismic design, which is a plot of the peak acceleration response of a series of single degree of freedom system of varying natural frequency with a given damping ratio, generally expressed by ratios of the responses to peak ground motion accelerations

2.1.16 dynamic method

method to analyze seismic response of structures based on the theory of structural dynamics

2.1.17 time history analysis method

method to analyze seismic response in whole time history by integrating the governing motion equation of structure with recorded accelerogram as seismic input

2.1.18 mode decomposition method

method to analyze seismic response of the structure, in which the total seismic response of the structure is obtained by superposition of seismic response of each mode. It is called the mode decomposition time history analysis method, when the time history analysis is used to obtain the seismic response of each mode. It is called the mode decomposition response spectrum method, when the response spectrum is used to obtain the seismic response of each mode

2.1.19 square root of the sum of the squares (SRSS) method

mode superposition method taking the square root of the sum of the squares of the seismic response of each mode as the total seismic response

2.1.20 complete quadric combination (CQC) method

mode superposition method for defining seismic response of structure as a square root of the sum of quadric terms of various mode seismic effects and coupling terms

2.1.21 seismic hydrodynamic pressure

dynamic pressure of water on structure caused by earthquake

2.1.22 seismic earth pressure

dynamic pressure of soil mass on structure caused by earthquake

2.1.23 quasi-static method

static analysis method taking the product of gravity action, ratio of design seismic acceleration to gravity acceleration and specified dynamic distribution coefficient as the design seismic action

2.1.24 seismic effect reduction factor

reduction factor for seismic effects introduced due to simplification in analysis method

2.1.25 natural vibration period

time interval for the structure to complete a free vibration cycle in a certain vibration mode. The natural vibration period corresponding to the first vibration mode is called the fundamental period

2.2 Symbols
2.2.1 Actions and Effects

a_h Representative value of horizontal design seismic acceleration

a_v Representative value of vertical design seismic acceleration

β Design response spectrum

ξ Seismic effect reduction factor

F_E Representative value of seismic active earth pressure

G_E Characteristic value of structure total gravity action

E_i Representative value of horizontal seismic inertial force exerting on mass point i

$P_w(h)$ Representative value of seismic hydrodynamic pressure at the depth h

F_0 Representative value of total seismic hydrodynamic pressure on water-contact face per unit width of structure

α_i Dynamic distribution coefficient of seismic inertial force of mass point i

g Gravitational acceleration $g=9.81$ m/s^2

2.2.2 Material Properties and Geometric Parameters

f_k Characteristic value of material property

a_k Characteristic value of geometric parameter

N Blow count of standard penetration test

N_{cr} Critical blow count

ρ_w Characteristic value of water mass density

v_p Characteristic value of compression wave velocity

v_s Characteristic value of shear wave velocity

K_u Characteristic value of stiffness coefficient of tunnel foundation per unit length at axial direction

K_v Characteristic value of stiffness coefficient of tunnel foundation per unit length perpendicular to axial direction

2.2.3 Limit State Design with Partial Factor

S Structure action effect

R Structure bearing capacity

γ_0 Structural importance factor

ψ Design situation factor

E_h Representative value of seismic action

G_h Characteristic value of permanent action

Q_h Characteristic value of variable action

γ_G Partial factor for permanent action

γ_Q Partial factor for variable action

γ_d Structural factor, safety margin introduced to consider the non-random uncertainty under limit condition of bearing capacity

γ_m Partial factor for material property

2.2.4 Structure Dynamic Characteristics

λ_m Mass ratio of appurtenant structure to main structure

λ_f Fundamental frequency ratio of appurtenant structure to main structure

T_g Characteristic period

T Natural vibration period of structure

3 Basic Requirements

3.0.1 The seismic fortification class of hydraulic structures shall be determined based on their importance and basic seismic intensity on their sites as per Table 3.0.1.

Table 3.0.1 Seismic fortification classification

Seismic fortification class	Grade of structure	Site basic intensity
A	Water-retaining and important water release structure of Grade 1	≥ VI
B	non-water-retaining structure of Grade 1 and water-retaining structure of Grade 2	
C	non-water-retaining structure of Grade 2 and structure of Grade 3	≥ VII
D	structure of Grade 4 and 5	

NOTE Important water release structures refer to those that might endanger the safety of retaining structures in case of failure.

3.0.2 The seismic fortification class of hydraulic structures shall be represented in terms of design intensity and horizontal design PGA on flat ground surface, and shall be determined as follows:

1 For the hydraulic structures of which the seismic fortification classes are determined in accordance with GB 18306, *Seismic Ground Motion Parameter Zonation Map of China*. For ordinary projects, the value of the PGA on their sites shall be taken from the map as the representative value of the design horizontal PGA after site category adjustment, and the corresponding basic seismic intensity is taken as the design intensity. For hydraulic structures of Class A seismic fortification, their design intensity shall be 1 intensity level higher than the basic intensity, and the representative value of the design horizontal PGA shall be doubled accordingly.

2 For projects whose seismic fortification criteria are based on site-specific seismic safety evaluation, the exceeding probability of the representative values of horizontal design PGA on the flat rock foundation surface shall be $P_{100} = 0.02$ in 100 years for water-retaining structures and important water-releasing structures with seismic fortification Class A. An exceeding probability in 50 years

P_{50} shall be 0.05 for non-water-retaining structures with seismic fortification Class B. An exceeding probability in 50 years P_{50} shall be 0.10 for hydraulic structures with other seismic fortification classes than A, and the corresponding PGA shall not be lower than that specified in the map.

3   For hydraulic structures with seismic fortification Class A whose design seismic parameters shall be provided by the site-specific seismic safety evaluation, a special demonstration on safety margin under the maximum credible earthquake (MCE) shall be carried out on disaster prevention of the uncontrolled release of reservoir in addition to the seismic design under design PGA. A special report shall be documented. The MCE of the site shall be determined by the deterministic method or the probabilistic method with an exceeding probability of 0.01 within 100 years of the reference period.

4   When the grade of water-retaining structure is raised from Grade 2 to Grade 1 due to dam height or seismic geological conditions, seismic design shall be carried out with design PGA of exceeding probability $P_{50} = 0.10$. Furthermore, a special demonstration on safety margin under PGA of exceeding probability $P_{100} = 0.05$ shall be carried out on disaster prevention of the uncontrolled release of reservoir.

5   In a special demonstration, relevant site-specific design response spectrum should be determined based on scenario earthquake corresponding to horizontal design PGA, and artificial synthetic time histories of seismic ground acceleration are generated. For the seismic effect analysis of structures with strong nonlinearity, frequency non-stationary influence of the ground motion should be studied. When the distance from the seismogenic fault to the site is less than 30 km and its inclination angle is less than 70°, hanging wall effect should be considered. When the distance is less than 10 km and magnitude is over 7.0, the fracture process of seismogenic fault as a near-fault strong area source should be studied to generate directly the random time histories of ground motion, and then to select the time histories with the peak period of evolutionary spectrum closest to the fundamental period of structure.

6   Seismic actions need not be considered during construction.

3.0.3   For new reservoirs with the dam higher than 100 m and storage capacity larger than 500 million m³, an evaluation of reservoir earthquake

shall be conducted. For reservoirs with potential reservoir earthquake of magnitude higher than 5 or epicentral intensity higher than Ⅶ, a reservoir earthquake monitoring network shall be established and put into operation at least one year prior to the initial impoundment.

3.0.4 The seismic design for hydraulic structures includes seismic calculation and measures for earthquake resistance, which shall meet the following requirements:

1. Select the region, site and structure type favorable for seismic resistance according to the seismic requirements.

2. Prevent stability failure of structure foundation and slope.

3. Select safe and cost-effective structures and measures for earthquake resistance.

4. Propose the construction quality control measures meeting the seismic safety requirements in design documents.

5. Provide water release structures that can lower the reservoir level as quickly as possible if necessary.

6. Conduct seismic designs for non-structural parts, appurtenance electromechanical equipment and their connections to main structures in hydraulic structures such as sluice, intake tower and shiplift.

3.0.5 The requirements for emergency plan to prevent and mitigate earthquake hazard shall be proposed in design for hydraulic structures with seismic requirements.

3.0.6 Dynamic model test should be conducted for dams of Class A fortification with design intensity of Ⅷ and above, and with height more than 150 m.

3.0.7 The seismic monitoring array design for strong-motion observation should be proposed for Grade 1 dam with design seismic intensity of Ⅶ and above, and Grade 2 dam with design seismic intensity of Ⅷ and above.

4 Site, Foundation and Slope

4.1 Site

4.1.1 In site selection for a hydraulic structure, a comprehensive evaluation shall be performed in terms of tectonic activity, the stability of site foundation and slope, and the risk of secondary disasters, etc., based on engineering geological and hydro-geological exploration and seismicity investigation. The site areas shall be classified into four categories: favorable, ordinary, unfavorable and hazardous according to Table 4.1.1. Favorable or ordinary site area for seismic design should be selected, and unfavorable and hazardous site area should be avoided. A thorough seismic safety evaluation must be conducted for a dam constructed in unfavorable and hazardous site area.

Table 4.1.1   Identification of site areas

Site area category	Tectonic activity	Stability of site foundation and slope	Secondary disaster
Favorable	No active fault within 25 km around the site, with basic intensity of Ⅵ	Good	Low
Ordinary	No active fault within 5 km around the site, with basic intensity of Ⅶ	Fair good	Fair low
Unfavorable	There are active faults shorter than 10 km within 5 km around the site; There are seismogenic structures with a magnitude less than 5. With basic intensity of Ⅷ	Fair poor	Fair high
Hazardous	There are active faults longer than 10 km within 5 km around the site; There are seismogenic structures with a magnitude larger than 5; With basic intensity of Ⅸ	Poor	High

4.1.2 The site soils after excavation and treatment for a hydraulic structure should be classified according to the shear wave velocity of soil layers shown in Table 4.1.2.

Table 4.1.2   Identification of site soil

Type	Shear wave velocity v_s(m/s)	Descriptions and features
Hard rock	$v_s > 800$	Hard, relatively hard sound rocks
Soft rock and hard soil	$800 \geq v_s > 500$	Fractured and relatively fractured, or soft and relatively soft rocks; dense sandy gravels
Moderately hard soil	$500 \geq v_s > 250$	Moderate-dense and slight-dense sandy gravels; dense coarse sand and medium sand; hard clay or silt
Moderately soft soil	$250 \geq v_s > 150$	Slight-dense gravels, coarse, medium and fine sand and silty sand; ordinary clay and silt
Soft soil	$v_s \leq 150$	Muck; mucky soil; loose sandy soil; miscellaneous fill

NOTES:

1. v_s refers to shear wave velocity of site soil. In the case of multi-layer site soil, equivalent shear wave velocity of site soil beneath the foundation is calculated with the formula $v_s = d_0 \Big/ \sum_{i=1}^{n}(d_i / v_{si})$. Where, d_0 is the overburden thickness (m); d_i is the thickness of the ith layer of the site soil (m); v_{si} is the shear wave velocity of the ith soil layer (m/s); and n is layer number of the site soil.

2. The overburden thickness d_0 shall generally equal the depth from the ground or foundation surface to the top of the layer, whose shear wave velocity is more than 500 m/s and the shear wave velocity of underlying layers is not less than 500 m/s, or whose depth is more than 5 m and shear wave velocity is 2.5 times more than that of overlying soil layer and the shear wave velocity of itself and underlying layers is not less than 400 m/s. The boulders and lenticles with a shear wave velocity greater than 500 m/s shall be deemed as surrounding soil layer. The hard rock layer intercalated in soil shall be considered as rigid body and its thickness shall be deducted from the overburden thickness.

4.1.3   Sites shall be classified into five categories, namely I_0, I_1, II, III, and IV, according to type of site soil and overburden thickness, as shown in Table 4.1.3.

Table 4.1.3   Identification of site categories

Type	Overburden thickness d_0(m)						
	0	$0 < d_0 \leq 3$	$3 < d_0 \leq 5$	$5 < d_0 \leq 15$	$15 < d_0 \leq 50$	$50 < d_0 \leq 80$	$d_0 > 80$
Hard rock	I_0	—					
Soft rock and hard soil		I_1	—				
Moderately hard soil			I_1		II		
Moderately soft soil			I_1	II		III	
Soft soil			I_1	II		III	IV

4.2 Foundation

4.2.1 In the seismic design of foundation for hydraulic structures, the type, load, hydraulic and operation conditions of superstructures, as well as engineering geological, hydrological conditions of foundation and bank slope shall be considered comprehensively.

4.2.2 For the foundation and bank slope of water-retaining structures, such as dam and sluice, the criteria on stability of earthquake liquefaction, earthquake subsidence of weak clay and seepage deformation under design seismic action shall be met. The deformation endangering the structures shall be avoided.

4.2.3 For weak discontinuities in foundation and bank slope of hydraulic structures, such as fractures, cracked zones, dislocation zone, and specially, mud-interbedded layer and argillization-liable rock layers with a low dip, the stability and deformation under design seismic action shall be verified according to their occurrence, buried depth, boundary conditions, seepage, physical and mechanical properties and design seismic intensity of structures. Seismic measures shall be taken if necessary.

4.2.4 For seepage control system and its connections, drainage and filters in hydraulic structure foundation and slope, effective measures shall be taken to prevent hazardous cracks or seepage damage under earthquakes.

4.2.5 For heterogeneous foundations, of which the material properties and thickness vary greatly in horizontal direction, measures shall be taken to prevent large differential settlement, sliding and concentrated seepage, and to improve the capacity of superstructure to tolerate differential settlement of the foundation.

4.2.6 Liquefaction of soil layer in foundation shall be identified according to GB 50287, *Code for Hydropower Engineering Geological Investigation*.

4.2.7 For potential liquefaction soil layers in foundation, the following seismic measures may be taken according to the specific conditions of the project:

1 Replace the liquefiable soil layers with non-liquefiable soil.

2 Use the artificial compaction strengthening methods, including vibroflotation and strong ramming, etc.

3 Adopt counter weight and drainage measures.

4 Adopt compound foundation like vibration-compacted stone column, or foundation with piles driven into the non-liquefiable soil layer underlying the liquefiable soil layer.

5 Confine the liquefiable foundation soil by continuous concrete walls or other measures.

4.2.8 For soft clay layers in the foundations of a hydraulic structure with a seismic fortification Class A or B, the special seismic test and analysis shall be carried out. Generally, foundation soil is identified as a soft clay layer if meeting any of the following criteria:

1 Liquefaction index $I_L \geq 0.75$.

2 Unconfined compressive strength $q_u \leq 50$ kPa.

3 Blow count of standard penetration test $N \leq 4$.

4 Sensitivity $S_t \geq 4$.

4.2.9 For soft clay layers in foundation, the following seismic measures may be taken according to the type of structures and specific conditions:

1 Remove or replace soft clay in the foundation.

2 Strengthen the layers with the preloading method.

3 Adopt the counter weight and sand well drain or plastic drainage board.

4 Set the pile foundation or composite foundation such as vibro-replacement stone column.

4.3 Slope

4.3.1 The distribution of unstable slopes under design seismic action shall be identified when complicated rock mass structures, weak discontinuities or unfavorable combinations of mud-interlayer exist within the site of a hydraulic structure. The potential hazard shall be analyzed and treatment measures shall be proposed.

4.3.2 The design intensity and representative value of design seismic acceleration of a slope shall be demonstrated based on the seismic fortification class of the relevant hydraulic structures, the relations between the slope and the hydraulic structures, and the failure impact of the slope on the hydraulic structures.

4.3.3 The rigid limit equilibrium method may be adopted for the calculation of slope seismic stability. Dynamic amplification effect of slope seismic inertial force may not be considered and static shear strength can be adopted in dynamic stability analysis.

4.3.4 The seismic analysis and the requirement of safety factor for slopes shall comply with DL/T 5353, *Design Specification for Slope of Hydropower* *and Water Conservancy Project.*

4.3.5 For important high slopes with complicated geological conditions, special studies shall be conducted based on dynamic analysis. The deformation and seismic stability safety of the slope shall be analyzed comprehensively on seismic effects such as the displacements, residual displacements or opening of sliding plane of the slopes.

5 Seismic Action and Seismic Calculation

5.1 Ground Motion Components and Combination

5.1.1 Generally, only horizontal seismic actions may be considered for hydraulic structures except aqueducts.

5.1.2 For water retaining structures of Grade 1 and 2, such as embankment dams and gravity dams with design intensity of Ⅷ or Ⅸ, and for long cantilevered, large-span or tall hydraulic concrete structures, both horizontal and vertical seismic actions shall be taken into account. Generally, the representative value of vertical design seismic acceleration may be taken as 2/3 the representative value of the horizontal design seismic acceleration, but shall be the representative value of the horizontal design seismic acceleration for near-source earthquakes.

5.1.3 For arch dams with special types such as severely asymmetric or hollow ones, and for Grade 1 and 2 double-curvature arch dams with design intensity of Ⅷ or Ⅸ, the vertical seismic action effects should be studied specially.

5.1.4 Generally, for embankment dams and concrete gravity dams, only the horizontal seismic actions along the stream direction may be considered in the seismic design. For blocks of gravity dam on steep slopes of river banks and for important embankment dams, the horizontal seismic actions along cross-stream direction should be considered.

5.1.5 For concrete arch dams and sluices, the horizontal seismic actions both stream and cross-stream direction shall be considered.

5.1.6 For intake towers, frames on the top of sluices and others hydraulic concrete structures with similar stiffness along the two principal axial directions, horizontal seismic actions along the two principal axial directions of structures shall be considered.

5.1.7 When the seismic action effects along orthogonal directions are calculated simultaneously by the mode decomposition method, the overall seismic action effects may be taken as the square root of the sum of squares (SRSS) of seismic action effects in each direction.

5.2 Classification of Seismic Actions

5.2.1 Generally, seismic actions to be considered in the seismic calculation of hydraulic structures shall include the inertial force of dead weight of structures and facilities, dynamic earth pressure and hydrodynamic pressure, as well as dynamic pore water pressure.

5.2.2 Hydrodynamic pressure shall be considered in seismic analysis for concrete face rockfill dams (CFRDs) and may be ignored for other embankment dams.

5.2.3 Seismic effect on wave pressure, seepage and uplift pressure may be ignored.

5.2.4 Generally, seismic effect on silt pressure may be ignored, but the water depth in front of a structure shall include silt deposit depth in hydrodynamic pressure calculation; if a high dam has an extremely deep silt deposit, the seismic effect on the silt pressure shall be studied specially.

5.3 Design Response Spectra

5.3.1 For hydraulic structures with seismic fortification Class A requiring site-specific seismic safety evaluation, the site-specific design response spectrum stipulated in Item 5 of Article 3.0.2 of this code shall be adopted as the design response spectrum. For other structures, standard design response spectrum shall be adopted as the horizontal and vertical design response spectrum.

5.3.2 Standard design response spectrum shall be Figure 5.3.2.

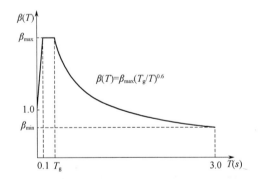

Figure 5.3.2 Standard design response spectrum

5.3.3 The representative value of maximum value of standard design response spectrum β_{max} for various hydraulic structures shall be taken in accordance with Table 5.3.3.

Table 5.3.3 Representative value of maximum value of standard design response spectrum β_{max}

Structure type	Embankment dam	Gravity dam	Arch dam	Sluice, intake tower and other structures and slopes
β_{max}	1.60	2.00	2.50	2.25

5.3.4 The representative value of lower limit of standard design response spectrum β_{mim} shall not be less than 20 % of that of maximum value of standard design response spectrum.

5.3.5 The characteristic periods of standard design response spectrum T_g for different site categories may be selected according to the site location specified in GB 18306, *Seismic Ground Motion Parameter Zonation Map of China*, and adjusted in accordance with Table 5.3.5.

Table 5.3.5 Adjustment for characteristic period of standard design response spectrum of site

Characteristic period of basic response spectrum of site category Ⅱ	Site category				
	I₀	I₁	Ⅱ	Ⅲ	Ⅳ
0.35 s	0.20 s	0.25 s	0.35 s	0.45 s	0.65 s
0.40 s	0.25 s	0.30 s	0.40 s	0.55 s	0.75 s
0.45 s	0.30 s	0.35 s	0.45 s	0.65 s	0.90 s

5.4 Combination of Seismic Action and Other Actions

5.4.1 Generally, upstream pool level in the seismic calculation of hydraulic structures may be taken as the normal pool level; for overyear regulating reservoir, it may be lower than the normal pool level after demonstration.

5.4.2 For upstream slope of an embankment dam, a frequently occurred pool level most unfavorable to its seismic stability shall be adopted in the seismic calculation according to the operation condition. If necessary, seismic action shall be considered for the condition of frequent pool level drop.

5.4.3 For seismic strength checking of important arch dam and sluice, an additional calculation should be done for the combination of seismic action and frequent low pool level.

5.5 Structural Modeling and Calculation Method

5.5.1 The structural modeling of seismic action effect for various hydraulic structures shall be the same as that stipulated in their corresponding design codes.

5.5.2 For embankment dams in narrow valley and integral gravity dams, the seismic calculation may be carried out by modelling the whole dam. For ordinary gravity dam, sluice and embankment dam, the seismic calculation may be conducted in terms of unit dam or sluice width or a single dam or sluice monolith.

5.5.3 In addition to the relevant provisions in this code, the calculation method of the seismic action effects for hydraulic structures shall be adopted according to the seismic fortification classes, as per Table 5.5.3.

Table 5.5.3 Calculation method of seismic action effect

Seismic fortification class	Calculation method of seismic action effect
A	Dynamic method, and for embankment dams quasi-static method may be also adopted
B, C	Dynamic method or quasi-static method
D	Quasi-static method or only seismic measures

5.5.4 For linear elasticity analysis of hydraulic structures, the mode decomposition response spectrum method or the mode decomposition time history analysis method may be adopted for calculating seismic action effects, in which only the elasticity influences of foundation are taken into account. The representative value of the design PGA shall be taken in accordance with Article 5.3.1 or Article 3.0.2 of this code. The damping ratios for various hydraulic structures may be taken as follows: 20 % for embankment dams, 5 % for arch dams, 10 % for gravity dams, 7 % for sluices, intake towers and other structures. For damping ratio of slope, special studies shall be carried out.

5.5.5 For concrete gravity dams and arch dams with seismic fortification Class A, in the special seismic analysis, the following factors should be taken into account in the dynamic analysis model: dynamic interaction of structure-foundation-reservoir system, the mass of foundation, rock type and geological conditions of near-field foundation, the far-field radiation damping, non-uniform ground motion, material nonlinearity of dam concrete and near-field foundation, opening and sliding of arch dam transverse joints during earthquakes. The massless foundation model may be used to consider the dynamic interaction between structure and foundation for other hydraulic structures when the seismic action effects are calculated by dynamic method.

5.5.6 When the mode decomposition response spectrum method is used, seismic action effects of various modes may be combined by the SRSS method. If the ratio of the absolute value of the frequency difference between two modes to the minor frequency is less than 0.1, seismic action effects should be combined by the CQC method.

$$S_E = \sqrt{\sum_i^m \sum_j^m \rho_{ij} S_i S_j} \qquad (5.5.6\text{-}1)$$

$$\rho_{ij} = \frac{8\sqrt{\zeta_i\zeta_j}(\zeta_i + \gamma_\omega\zeta_j)\gamma_\omega^{3/2}}{(1-\gamma_\omega^2)^2 + 4\zeta_i\zeta_j\gamma_\omega(1+\gamma_\omega^2) + 4(\zeta_i^2 + \zeta_j^2)\gamma_\omega^2} \qquad (5.5.6\text{-}2)$$

where

S_E is the seismic action effect;

S_i, S_j are the seismic action effects of ith and jth modes, respectively;

m is the number of modes used for calculation;

ρ_{ij} is the correlation coefficient of ith and jth modes;

ζ_i, ζ_j are the damping ratios of ith and jth modes, respectively;

γ_ω is the circular frequency ratio $\gamma_\omega = \omega_j / \omega_i$;

ω_i, ω_j are the circular frequencies of ith and jth modes, respectively.

5.5.7 Higher order modes with contribution of no more than 5 % to seismic action effects may be ignored. When the lumped mass model is used, number of lumped masses should not be less than four times that of modes used in the calculation of seismic action effects.

5.5.8 When the time history analysis method is used for calculating seismic action effects, design response spectrum with 5 % damping ratio shall be taken as target spectrum. At least three sets of artificial accelerograms shall be generated as the design seismic accelerograms, and the correlation coefficient between components of each set of those accelerograms shall not be larger than 0.3. The peak values of the design seismic accelerograms shall be adopted according to Article 3.0.2 of this code. The calculated results with various seismic accelerograms shall be comprehensively analyzed to determine the seismic action effects for the design.

5.5.9 When the quasi-static method is used for the calculation of seismic action effect, the representative value of horizontal seismic inertial force acting on mass point i along height of the structure shall be calculated by the following equation:

$$E_i = a_h \xi G_{Ei} \alpha_i / g \qquad (5.5.9)$$

where

E_i is the representative value of horizontal seismic inertial force acted on mass point i;

ξ is the reduction factor for seismic action effect, which shall be taken as 0.25 unless otherwise specified;

G_{Ei} is the characteristic value of gravity action concentrated on mass point i;

α_i is the dynamic distribution coefficient of seismic inertial force of mass point i, which shall be taken in accordance with the relevant provisions of this code;

g is the gravity acceleration.

5.6 Dynamic Properties of Concrete and Foundation Rock Mass

5.6.1 For mass concrete hydraulic structures with seismic fortification Class A, dynamic properties of the concrete shall be determined by test.

5.6.2 For mass concrete hydraulic structures whose dynamic properties of concrete are not determined by test, the characteristic values of concrete dynamic strength may be 120 % of static ones, and the corresponding partial factors for material properties may be taken as 1.5; the characteristic values of dynamic elastic modulus may be 150 % of static ones; the characteristic value of dynamic tensile strength may be 10 % of that of the dynamic compressive strength.

5.6.3 For seismic stability calculation of concrete hydraulic structures, dynamic deformation modulus of foundation rock mass may be taken as static deformation modulus; when the dynamic method is adopted for calculation of seismic action effect, the characteristic values of dynamic shear strength parameters of foundation rock mass as well as the interface between rock foundation and concrete may be taken as the static ones. When the quasi-static method is adopted to calculate seismic action effects, the characteristic values of dynamic shear strength parameters of foundation rock mass as well as the interface between rock foundation and concrete shall be the mean values of static shear strength parameters.

5.7 Seismic Design for Ultimate Limit States with Partial Factors

5.7.1 The seismic strength and stability under the most unfavorable combinations considering both static and dynamic actions for various hydraulic structures shall satisfy the design expression (5.7.1) for ultimate limit states. If not, special studies shall be conducted.

$$\gamma_0 \psi S(\gamma_G G_K, \gamma_Q Q_k, \gamma_E E_k, a_k) \leq \frac{1}{\gamma_d} R\left(\frac{f_k}{\gamma_m}, a_k\right) \tag{5.7.1}$$

where

γ_0 is the importance factor for structure, which shall be taken in accordance with GB 50199, *Unified Standard for Reliability Design of Hydraulic Engineering Structures*;

ψ is the design situation factor, taken as 0.85;

$S(\cdot)$ action effect function of structure;

γ_G partial factor for permanent action;

G_k characteristic value of permanent action;

γ_Q partial factor for variable action;

Q_k characteristic value of variable action;

γ_E partial factor for seismic action, taken as 1.0;

E_k representative value of seismic action;

a_k characteristic value of geometric parameter;

γ_d structural factor for ultimate limit states;

$R(\cdot)$ resistance function of structure;

f_k characteristic value of material property;

γ_m partial factor for material property.

5.7.2 Limit states to be checked and corresponding structural factors for various hydraulic structures under seismic actions shall follow the relevant stipulations in this code.

5.7.3 Partial factors for and characteristic values of static actions combined with seismic action shall be assigned in accordance with corresponding design codes for various structures. When partial factors for actions and resistance are not stipulated in these design codes, or a reduction factor for seismic action effect is introduced in the seismic checking, partial factors may be taken as 1.0.

5.7.4 In the seismic design of reinforced concrete structural elements, the seismic checking of their sectional bearing capacity shall be conducted in accordance with DL/T 5057, *Design Specification for Hydraulic Concrete Structures* with the seismic action effect determined by this code. When the dynamic method is used to calculate seismic action effects, the reduction factor for seismic action effect shall be taken as 0.35. When the quasi-static method is used to calculate seismic action effects of reinforced concrete structural elements, the reduction factor for seismic action effect shall be taken as 0.25 in seismic inertial force calculation.

5.7.5 The partial factors for dynamic material properties may take the static ones.

5.8 Seismic Calculation for Appurtenant Structure

5.8.1 In the calculation of seismic action effects for appurtenant structures of a hydraulic structure, if the mass ratio λ_m and fundamental frequency ratio λ_f of appurtenant structures to main structure satisfy either of the following conditions, coupling analysis of appurtenant structures and main structure need not be performed:

1 $\lambda_m < 0.01$.

2 $0.01 \leq \lambda_m \leq 0.1$ and $\lambda_f \leq 0.8$ or $\lambda_f \geq 1.25$.

5.8.2 For an appurtenant structure not considering coupling analysis, the seismic input in the seismic action effect calculation may be taken as the acceleration at the connection with the main structure.

5.8.3 For an appurtenant structure and main structure not considering coupling analysis, their connections may be regarded rigid, and the mass of the appurtenant structure shall be considered as an added mass of the main structure.

5.9 Seismic Earth Pressure

5.9.1 The representative value of seismic active earth pressure may be calculated by Formula (5.9.1-1) and shall take the greater value of the calculated results with plus or minus signs in Formula (5.9.1-1).

$$F_{\mathrm{E}} = \left[q_0 \frac{\cos\psi_1}{\cos(\psi_1 - \psi_2)} H + \frac{1}{2}\gamma H^2 \right] (1 \pm \xi a_v / g) C_e \qquad (5.9.1\text{-}1)$$

$$C_e = \frac{\cos^2(\varphi - \theta_e - \psi_1)}{\cos\theta_e \cos^2\psi_1 \cos(\delta + \psi_1 + \theta_e)(1 + \sqrt{Z})^2} \qquad (5.9.1\text{-}2)$$

$$Z = \frac{\sin(\delta + \varphi)\sin(\varphi - \theta_e - \psi_2)}{\cos(\delta + \psi_1 + \theta_e)\cos(\psi_2 - \psi_1)} \qquad (5.9.1\text{-}3)$$

where

F_{E} is the representative value of seismic active earth pressure;

q_0 is the load per unit length on earth surface;

ψ_1 is the included angle of retaining wall surface with vertical plane;

ψ_2 is the included angle of earth surface with horizontal plane;

H is the height of earth;

γ is the characteristic value of gravitational density of earth;

φ is the internal friction angle of earth;

θ_e is the seismic coefficient angle, $\theta_e = \arctan\dfrac{\xi a_h}{g \pm \xi a_v}$;

δ is the friction angle between retaining wall surface and earth;

ξ is the reduction factor for seismic action effect; when the dynamic method is used for the calculating of seismic action effect, it shall be taken as 1.0; when the quasi-static method is used, it shall be usually taken as 0.25, and 0.35 for reinforced concrete structure.

5.9.2 The seismic passive earth pressure shall be determined through special study.

6 Embankment Dam

6.1 Seismic Calculation

6.1.1 The seismic calculation of embankment dams shall include seismic stability calculation, residual deformation calculation, safety evaluation of impervious body, discrimination of liquefaction, etc. Comprehensive evaluation of seismic safety shall be performed in combination with seismic measures.

6.1.2 For seismic stability calculation of embankment dams, the quasi-static method generally shall be adopted to calculate seismic action effect. FEM shall also be adopted to analyze the seismic effect on dam body and foundation to comprehensively evaluate the seismic stability in one of the following conditions:

1 With a design seismic intensity of Ⅶ and a dam height over 150 m.

2 With a design seismic intensity of Ⅷ or Ⅸ and a dam height over 70 m.

3 With an overburden layer thicker than 40 m or liquefiable soils existing in dam foundation.

6.1.3 When the quasi-static method is adopted to calculate seismic action effects and seismic stability for embankment dams, the slip-circle method considering effect of inter-slice forces should be adopted in compliance with Article 5.7.1 of this code and refer to the calculation formulae in Appendix A of this code. The sliding wedge method may be adopted for foundations with thin soft clay layers, and dams with thin inclined clay core or thin clay core.

6.1.4 When the quasi-static method is adopted to calculate seismic action effects and seismic stability for embankment dams, the dynamic distribution coefficients of mass point i shall be taken in compliance with Table 6.1.4. In this table, α_m is taken as 3.0, 2.5 and 2.0 for design intensity Ⅶ, Ⅷ and Ⅸ, respectively.

Table 6.1.4 Dynamic distribution coefficient of seismic inertial force in embankment dam

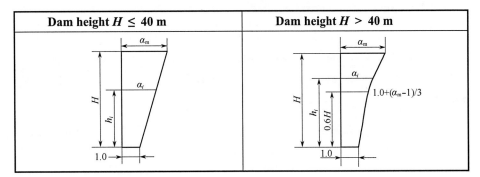

6.1.5 When the quasi-static method is adopted to calculate the seismic action effects and stability for Grade 1 and 2 embankment dams, the dynamic shear strength of soil should be determined through dynamic tests. If the tested dynamic strength is higher than the corresponding static strength, the static strength shall apply.

When dynamic test data is unavailable for non-liquefiable soil like cohesive soil and dense sandy gravel, the static effective shear strength may be adopted; and for coarse cohesionless soil like rockfill and sandy gravels, the non-linear static shear strength considering confining pressure influences should be adopted.

6.1.6 Dynamic analysis of seismic action effects on embankment dams with FEM should be conducted in compliance with the following requirements:

 1   Calculate the initial stress state before earthquake with non-linear stress-strain relations of materials.

 2   Determine the characteristic parameters of dynamic deformation, dynamic residual deformation, and dynamic strength of dam materials through material dynamic tests, in combination of engineering analogy.

 3   Calculate seismic response with dynamic non-linear stress-strain relations of dam materials.

 4   Analyze seismic stability along potential sliding plane based on seismic action effects and calculate residual deformation of dam body caused by earthquake.

 5   Comprehensively evaluate the seismic safety based on seismic response analysis in such aspects as slope stability, residual deformation, safety of impervious body, and liquefaction discrimination, in accordance with Article 6.1.11 of this code.

6.1.7 The material samples for dynamic tests shall be representative and test conditions shall reflect compactness state and consolidated stress state of dam body and foundation soil. Dynamic characteristic parameters of dam materials, if available, should be determined through lab tests combined with in-situ tests.

6.1.8 In the calculation of dam residual deformation, the influence of residual volumetric strain and residual shear strain should be taken into account.

6.1.9 The hydrodynamic pressure of CFRD may be determined in accordance with Articles 7.1.12 to 7.1.14 of this code.

6.1.10 When the slice method considering inter-slice forces is adopted, the structural factor shall not be smaller than 1.2. When the slice method without considering inter-slice forces is adopted for seismic stability calculation, the structural factor shall not be smaller than 1.1.

6.1.11 Seismic safety evaluation based on dynamic calculation results should be conducted in compliance with the following requirements:

1. Comprehensively evaluate the seismic sliding stability of dam slope and its influences on overall seismic safety of dam according to the location, depth, and range of sliding circle, and duration and magnitude of limit exceedance of seismic stability.

2. Determine the distribution range of local shear failure or liquefaction failure of dam body, and evaluate the possibility of inducing global failure of dam.

3. Provide the magnitude and distribution of residual deformation of dam body, and comprehensively evaluate the seismic safety of dam and impervious body according to the maximum seismic settlement ratio and the unevenness of deformation.

6.1.12 In the seismic safety evaluation of embankment dams under MCE, finite element model reflecting the interaction of dam-foundation-reservoir system shall be built, and non-linear dynamic constitutive model and calculation parameters based on material dynamic tests shall be adopted, and the influences on stress and deformation of impervious body due to seismic residual deformation of dam body shall be taken into account. For important embankment dams specified in Article 3.0.6 of this code, a comprehensive evaluation shall be conducted based on dynamic calculation, model test and engineering analogy, to meet the fortification requirement of preventing uncontrolled release of reservoir.

6.2 Seismic Measures

6.2.1 For embankment dam in high intensity zone, straight axis or convex axis toward upstream should be adopted, and axis convex toward downstream or with broken or S-shape line should not be used.

6.2.2 When the design intensity is Ⅷ or Ⅸ, rockfill dam should be selected and impervious body should not be of rigid core. When a homogeneous dam is selected, an internal drainage system shall be established to lower the phreatic line.

6.2.3 The freeboard of an embankment dam in high intensity zone shall include the seismic surge height and earthquake-induced settlement, and it may be determined by the following principles:

1. Take a seismic surge height of 0.5 m to 1.5 m according to the design

intensity and water depth in front of the dam.

2 When the design intensity is Ⅶ, Ⅷ or Ⅸ, the additional settlement of dam body and foundation due to seismic actions shall be incorporated into the freeboard.

3 The possible surge resulted from earthquake-induced massive reservoir bank collapse or landslide, etc., shall be specially studied.

6.2.4 When the design intensity is Ⅷ or Ⅸ, the dam crest should be widened and the upper dam slope should be gentler. The dam slope toe may be strengthened with blanket or weights, upper dam slope may be strengthened with masonry revetment, and interior of upper dam slope may be strengthened with reinforcing bars, geosynthetics or concrete grids, etc.

6.2.5 The seismic performance of impervious body of embankment dam in seismic zones shall be properly improved, especially for such parts vulnerable to cracking during earthquake as upper dam body and crest, interfaces between dam body and bank slopes or other concrete structures. The interface between impervious body and bank slopes or concrete structures should not be too steep. Slope-change angle should not be too large and reverse slope or sudden slope change shall be avoided. The impervious body, upstream and downstream filters, and transition zones shall be appropriately thickened.

6.2.6 Well-graded embankment material with favorable dynamic properties and seepage stability shall be adopted. Uniform medium sand, fine sand, silt and silty soil should not be used for embankment in high intensity zone.

6.2.7 Compactive performance and compactness of cohesive soil, and dry density or porosity of rockfill shall be determined in accordance with DL/T 5395, *Design Specification for Rolled Earth-Rock Fill Dams* and DL/T 5016, *Design Specification for Concrete Face Rockfill Dams*. The upper limits stipulated should be taken when the design intensity is Ⅷ or Ⅸ.

6.2.8 For the compaction of cohesionless soil, the relative density of material above phreatic line shall not be lower than 0.75, and that below phreatic line shall be appropriately increased according to the design intensity. For sand-gravel material, if the content of coarse grain greater than 5 mm is less than 50 %, the relative density of fine materials shall meet the compaction requirements of cohesionless soil mentioned above, and the compact dry densities for various gravel contents, based on the above requirements, shall be used as filling criteria.

6.2.9 For Grade 1 and Grade 2 embankment dams, buried water-conduit should not be set under the dam. If unavoidable, reinforced concrete conduit or cast iron pipes shall be used and placed in the trench in rock foundation and the pipe shall lie below the bottom of dam body, and the conduit shall be wrapped up with concrete. Anti-seepage and sealing at the joints of conduit shall be reliable, and the control valve of the conduit shall be installed in the intake or the upstream side of imperious body. Reliable filter shall be arranged at conduit outlet and joints.

6.2.10 The liquefiable strata or soft clay foundation of embankment dams in seismic zone should be treated according to Article 4.2.7 or 4.2.9 of this code.

6.2.11 For CFRDs, the following seismic structural measures should also be taken:

1. Thicken the cushion zone and strengthen its connection with foundation and bank slopes. For steep bank slopes, the length of interface between cushion zone and bedrock should be appropriately increased and finer cushion material should be adopted.

2. Fill the vertical joints of facing at the middle of riverbed with asphalt coated board or other materials of desirable strength and flexibility.

3. Appropriately increase the reinforcement ratio for upper part of the facing at the middle of riverbed, especially along slope direction.

4. Make the construction joints of facing constructed in stages perpendicular to facing and provide double-layer reinforcement and stirrups in the vicinity of joints.

5. Use water-stop structures with good deformation performance but minimize their reduction to the facing cross section.

6. Appropriately increase the compaction density of rockfill material and pay special attention to compaction quality at abrupt changes of terrain.

7. Arrange internal drainage zone to ensure smooth drainage and fill the certain area at downstream dam slope with rockfill materials when filling the dam with soft rocks and sand-gravel materials.

7 Gravity Dam

7.1 Seismic Calculation

7.1.1 For gravity dams, seismic analyses of stress of dam body and overall stability against sliding along dam-foundation interface shall be carried out. For roller-compacted concrete (RCC) gravity dams, the seismic analysis of sliding stability along lift joints shall also be carried out.

7.1.2 The highest monolith of different dam sections may be selected for seismic analysis. For gravity dam with sound integrity, the analysis of whole dam should be carried out.

7.1.3 The dynamic method or quasi-static method may be adopted for seismic calculation of gravity dams. For gravity dams with seismic fortification Class A, or with seismic fortification Classes B and C with design intensity Ⅷ or higher, or with a dam height more than 70 m, the dynamic method shall be adopted.

7.1.4 The analysis of overall stability against sliding along dam-foundation interface for gravity dams and the analysis of stability against sliding along lift joints for RCC gravity dams shall be carried out by the shear strength formula for the rigid body limit equilibrium method. For the stability against deep-seated sliding, the rigid body limit equilibrium method based on the equal safety factor (equal-K method) shall be used. For gravity dams with complex foundation conditions, the nonlinear finite element analysis should be carried out as a supplement.

7.1.5 For stress analysis of gravity dams with a height more than 70 m, FEM analysis shall be conducted as supplement to the dynamic and static analysis using the cantilever method. For gravity dams with seismic fortification Class A, complex structure or complicated geological conditions, material nonlinearity shall also be considered in FEM analysis. For seismic calculation of gravity dams under MCE, FEM analysis shall be carried out considering material nonlinearity of dam body and foundation.

7.1.6 The mode decomposition method shall be adopted for the dynamic analysis of gravity dam. For gravity dam with seismic fortification Class A, a nonlinear FEM analysis shall be added.

7.1.7 Under design seismic action, the dynamic method is adopted in checking the strength and sliding stability of gravity dams along dam-foundation interface and RCC lift joints. When the cantilever method or FEM with equivalent stress treatment is adopted, the structural factors for compressive strength and tensile strength shall not be less than 1.3 and 0.7

respectively. The structural factor for sliding stability along dam-foundation interface or RCC lift joints shall not be less than 0.65, or shall be further assessed by time history analysis.

7.1.8 When the dynamic method is used to check the deep-seated sliding stability of a gravity dam, the parameters of rock mass shear strength shall take the static mean values, its partial factor shall be taken as 1.0, the structural factor for deep-seated sliding stability shall not be less than 1.4 or shall be comprehensively assessed by time history analysis.

7.1.9 The comprehensive evaluation of seismic stability against sliding along dam-foundation interface, RCC lift joints and deep-seated sliding plane shall be carried out by time history analysis in the following steps:

1. In each time step, structural factor for deep-seated sliding stability is calculated with the rigid body limit equilibrium method and a time history of structural factor is provided. The deep-seated sliding stability shall be evaluated based on the minimum structural factor in the time history.

2. If the minimum value of structural factor in the time history cannot meet the requirements of Article 7.1.7 or Article 7.1.8 of this code, the dam stability against sliding shall be comprehensively evaluated based on the duration and degree of limit exceedance.

7.1.10 When seismic safety demonstration for dams under MCE is required, the finite element model of dam-foundation-reservoir system with appropriate parameters shall be established for analysis, considering radiation damping effect of far-field foundation, material nonlinearities of dam concrete and near-field rock mass, etc. For important gravity dams specified in Article 3.0.6 of this code, a comprehensive evaluation based on calculation and model test results and engineering analogy shall be conducted to meet the seismic safety requirement of preventing uncontrolled release of reservoir.

7.1.11 When the quasi-static method is used for calculating seismic action effect on gravity dams, the representative values of horizontal seismic inertial force of various mass points shall be calculated in accordance with Article 5.5.9 of this code, in which, the dynamic distribution coefficients shall be determined by Formula (7.1.11).

$$\alpha_i = 1.4 \frac{1 + 4(h_i/H)^4}{1 + 4\sum_{j=1}^{n} \frac{G_{Ej}}{G_E}(h_j/H)^4} \quad (7.1.11)$$

where

> n is the total number of mass points for the calculation of dam body;
>
> H is the dam height, but for overflow dam it is taken up to the top of pier;
>
> h_i, h_j are the heights of mass points i and j, respectively;
>
> G_{Ej} is the characteristic value of gravity action concentrated on mass point j;
>
> G_E is the characteristic value of total gravity action of structures causing seismic inertial force.

7.1.12 When the quasi-static method is used to calculate the seismic action effects on a gravity dam, the representative value of seismic hydrodynamic pressure at water depth h shall be calculated by Formula (7.1.12-1).

$$P_w(h) = a_h \xi \psi(h) \rho_w H_0 \qquad (7.1.12\text{-}1)$$

where

> $P_w(h)$ is the representative value of seismic hydrodynamic pressure on vertical waterward face of dam at water depth h;
>
> $\psi(h)$ is the distribution coefficient of seismic hydrodynamic pressure at water depth h, which shall be taken in accordance with Table 7.1.12;
>
> ρ_w is the characteristic value of mass density of water;
>
> H_0 is the total water depth.

The representative value of the total seismic hydrodynamic pressure per unit width of dam face at water depth $0.54 H_0$ shall be calculated by Formula (7.1.12-2).

$$F_0 = 0.65 a_h \xi \rho_w H_0^2 \qquad (7.1.12\text{-}2)$$

Table 7.1.12 Distribution coefficient of hydrodynamic pressure on gravity dam

h/H_0	$\psi(h)$	h/H_0	$\psi(h)$
0.0	0.00	0.3	0.68
0.1	0.43	0.4	0.74
0.2	0.58	0.5	0.76

Table 7.1.12 *(continued)*

h/H_0	$\psi(h)$	h/H_0	$\psi(h)$
0.6	0.76	0.9	0.68
0.7	0.75	1.0	0.67
0.8	0.71		

7.1.13 The representative value of hydrodynamic pressure calculated by Formula (7.1.12-1) shall be multiplied by a reduction factor for the dam with sloping waterward face with an angle θ to horizontal plane:

$$\eta_c = \theta/90 \tag{7.1.13}$$

where

η_c is the reduction factor for hydrodynamic pressure;

θ is the acute angle between waterward face and horizontal plane (°).

When there is a break in the slope of the waterward face of dam and the height of the vertical portion below the water surface is equal to or greater than half of the water depth, the waterward face can be regarded approximately vertical, otherwise the line connecting the highest wetted point and the heel shall be regarded as the slope.

7.1.14 When the dynamic method is used, the hydrodynamic pressure corresponding to the unit seismic acceleration, calculated by Formula (7.1.14), may be converted into an added mass on dam face in the normal direction.

$$P_w(h) = \frac{7}{8} a_h \rho_w \sqrt{H_0 h} \tag{7.1.14}$$

7.1.15 When the quasi-static method is used to check the strength of dam body and sliding stability along dam-foundation interface, RCC lift joints, and deep-seated sliding planes, the structural factors for compressive strength and tensile strength shall not be less than 2.80 and 2.10, respectively, and the structural factor for sliding stability shall not be less than 2.70.

7.2 Seismic Measures

7.2.1 A gravity dam should adopt straight axis in the layout.

7.2.2 A gravity dam shall be simple in shape, avoiding abrupt changes in dam slope, and the slope break near dam crest should be curved. The dam crest should not be excessively inclined toward upstream. For the upper part of dam body, the weight should be reduced and the rigidity should be increased and the concrete should be upgraded or reinforced accordingly.

7.2.3 Appurtenant structures on dam crest shall be light, simple, well-integrated and minimized in height, and high tower structures should be avoided. The connections between the access bridge and the piers on overflow section as well as lateral rigidity of piers should be strengthened.

7.2.4 Weaknesses in foundation such as faults, fractures and weak intercalated layers shall be treated with engineering measures, and the bottom concrete should be upgraded and clay blanket should be provided at dam heel if necessary.

7.2.5 For the locations where the dam sections differ significantly along the dam axis or longitudinal topography and geology change abruptly, contraction joints shall be provided, and joint sealing and water-stop material with good flexibility should be selected.

7.2.6 For gravity dams with seismic fortification Class A and with a design acceleration greater than $0.2g$, contraction joints between dam monoliths should be provided with shear keys or grouting to improve dam integrity. Water-stop design for contraction joints shall be emphasized, and joint sealing and water-stop material with good flexibility should be selected.

7.2.7 Seismic weaknesses like orifice perimeter and connections between gate pier and weir surface of overflow dam shall be reinforced additionally.

8 Arch Dam

8.1 Seismic Calculation

8.1.1 Seismic calculation of an arch dam shall include the analysis of dam body stress and arch abutment stability under design earthquake. Deformation analysis of dam-foundation system shall also be made additionally when calculation under MCE is required.

8.1.2 Seismic calculation of an arch dam may be conducted with the dynamic method or the quasi-static method. The dynamic method shall be used to calculate seismic action effects on arch dams with seismic fortification Class A, with seismic fortification Class B or C but with a design intensity of VIII or higher, or with a height over 70 m.

8.1.3 For arch dams higher than 70 m, the stress analysis shall be conducted by FEM in addition to the static and dynamic trial-load methods. For arch dams with seismic fortification Class A, complex structure, or complicated geology, material non-linearity shall be considered in finite element analysis.

8.1.4 The mode decomposition method shall be adopted in the dynamic analysis of arch dams. For arch dam with seismic fortification Class A, a nonlinear FEM analysis shall be added.

8.1.5 Representative value of horizontal hydrodynamic pressure may be taken as 1/2 of the value calculated by Formula (7.1.14), where H_0 is the water depth of the calculated section. When the quasi-static method is adopted for analysis, the representative value of horizontal hydrodynamic pressure shall be multiplied by the dynamic distribution factor α_i and the seismic effect reduction factor ζ specified in Article 8.1.13 of this code. When the dynamic method is adopted, the hydrodynamic pressure under effect of unit horizontal acceleration may be converted to the added mass in normal direction on dam face.

8.1.6 When the dynamic method is used to check dam strength under design seismic actions, the structural factors for compressive and tensile strengths shall not be less than 1.30 and 0.70, respectively.

8.1.7 The stability of arch abutments under design seismic actions shall be calculated by shear strength formula of the rigid body limit equilibrium method.

8.1.8 Seismic stability calculation for arch abutments (including thrust block) under design seismic actions may be conducted according to the requirements below and Article 8.1.9 and Article 8.1.10 of this code. It may also be evaluated comprehensively by comparison of various methods.

1   After identifying the possible sliding rock blocks, the maximum value and direction of the thrust of arch abutments shall be determined according to the most unfavorable combination from the dynamic and static calculation of dam body.

2   The representative value of seismic inertial force of possible sliding rock blocks shall be calculated by Formula (5.5.9), where a_i is taken as 1.0. When the maximum thrust at arch abutments is determined by the dynamic method, the seismic effect reduction factor ξ of seismic inertial force of the rock shall be taken as 1.0, assuming that representative value of rock seismic inertial force and the maximum thrust at arch abutment occur simultaneously. For calculation of seismic inertial force of the rock block, the PGA in each direction shall be combined as follows:

   1) When PGA in cross-stream direction is taken as the design value, the PGAs in stream direction and vertical direction are taken as 1/2 of the design value.

   2) When PGA in stream direction is taken as the design value, the PGAs in cross-stream direction and vertical direction are taken as 1/2 of the design value.

   3) When PGA in vertical direction is taken as the design value, the PGAs in stream direction and cross-stream direction are taken as 1/2 of the design value.

3   The most unfavorable time-independent sliding mode shall be selected based on the geometrical features of the potential sliding rock blocks.

4   The influence of earthquake on seepage pressure variation in rock mass may be ignored.

8.1.9 When the dynamic method is used to check the stability of rock mass of arch abutment under design seismic actions, the shear strength parameters of the rock mass shall take the static mean values, the partial factor shall be taken as 1.0, and the structural factor for sliding stability shall not be less than 1.40; or the sliding stability of potential sliding rock mass in arch abutments shall be further assessed by time history analysis.

8.1.10 Comprehensive evaluation of seismic stability of potential sliding rock mass in arch abutment by time history analysis shall follow the procedures below:

   1   Under the actions of three components of design ground motion, the

resultant force time history of static and dynamic composite effect on the arch abutment is calculated by time history analysis and is exerted on potential sliding rock blocks together with the time history of the inertial force of rock mass ignoring dynamic amplification.

2   In each time step, the structural factor for stability of abutment rock mass is calculated with the rigid limit equilibrium method and the time history of structural factor is provided. The abutment seismic stability shall be evaluated according to the minimum structural factor in time history.

3   If the minimum value of structural factor in time history cannot meet the requirements in Article 8.1.9 of this code, the abutment sliding stability and its effects on overall dam safety shall be comprehensively evaluated based on the duration and degree of limit exceedance.

8.1.11   When seismic safety demonstration for dams under MCE is required, the influences of the contact nonlinearity of contraction joints on dam body and the slip surface of critical sliding rock mass, material nonlinearity of major weak zone in near-field foundation, and radiation damping effect of far-field foundation, shall be considered. For important arch dams specified in Article 3.0.6 of this code, a comprehensive evaluation based on calculation and model test results and engineering analogy shall be conducted to meet the seismic safety requirement of preventing uncontrolled release of reservoir.

8.1.12   When the seismic safety evaluation of arch dam is conducted according to Article 8.1.11 of this code, turning point on curves of deformation at typical locations on dam body or on foundation rock mass varying with the increase in input acceleration may be used to evaluate the safety of dam foundation system, and the ratio of the input acceleration at the turning point to the design seismic acceleration may be taken as the safety margin preventing uncontrolled release of reservoir.

8.1.13   When the seismic effects of arch dam are calculated by the quasi-static method, the representative values of horizontal seismic inertial forces of various mass points on different arch rings acting in normal direction shall be calculated according to Article 5.5.9 of this code, in which the dynamic distribution coefficients of dam crest and the lowest elevation of dam shall take 3.0 and 1.0, respectively, shall be interpolated vertically, and shall be uniform along arch rings.

8.1.14   When the strength of dam body and stability of arch abutments are checked with the quasi-static method, the structural factors for compressive and tensile strengths shall not be less than 2.80 and 2.10, respectively, and that for sliding stability shall not be less than 2.70.

8.2 Seismic Measures

8.2.1 The dam shape shall be selected reasonably to improve thrust direction of arch abutments and reduce tensile stress zones in upper and middle parts of dam and near the foundation under seismic action. For a double curvature arch dam, overhanging towards upstream side shall be checked, and the upper arch crown should be inclined properly toward downstream side.

8.2.2 The seismic stability of abutments shall be strengthened. Excessively large difference in mechanical properties and structure of rock mass between both banks shall be avoided and dam abutments shall not sit on a thin mountain ridge. Weak zones in rock foundation may be strengthened by grouting, concrete plug, local anchorage, supporting, etc. If necessary, the measures may be taken to thicken the dam body and deepen the excavation near the abutments of the top arch. To minimize seepage pressure in rock mass, curtain and drainage shall be provided in dam foundation. The pressure tunnel shall not be close to the dam abutment.

8.2.3 Detailing of the dam joints, in particular, waterstop, grouting temperature control and keys, shall be designed carefully. Waterstop with proper shape and material shall be adopted to accommodate repeated joint opening and closing during earthquake. If calculation results indicate that excessive contraction joint deformation under seismic action endangers the waterstop, dampers on crest and joint-crossing bars at upper dam section should be considered.

8.2.4 In high tensile stress zone of arch dam faces, especially the middle part of downstream face, higher strength concrete and seismic reinforcement may be provided. The dam crest with lighter weight and higher rigidity is preferred. Clay blanket, if necessary, may be provided at dam heel.

8.2.5 Light, simple and well-integrated appurtenant structures on the dam crest should be adopted, and the part extruding out of the dam body should be minimized. Arch thrust transferring structures should be provided between piers of the crest spillway. Connections on the crest, such as access bridge. shall be strengthened to prevent them from fall during earthquake.

9 Sluice

9.1 Seismic Calculation

9.1.1 The seismic calculation of a sluice shall include the check of seismic stability and structural strength. Seismic stability calculation for sluice chamber and side connections and their foundation shall be carried out; the seismic stress calculation of structural components of sluice, shall be carried out. Seismic design shall be conducted for nonstructural components, auxiliary electromechanical equipment and its connecting parts with the main structure.

9.1.2 The seismic action effect of a sluice may be calculated by the dynamic method or the quasi-static method. The dynamic method shall be adopted for Grade I and II sluices with the design seismic intensities of VIII and IX or seated on liquefiable soil.

9.1.3 When the quasi-static method is used to calculate the seismic action effect on a sluice, the representative values of horizontal seismic inertial forces of various mass points shall be calculated in accordance with Article 5.5.9 of this code, in which the dynamic distribution coefficient of seismic inertial force α_i shall be taken from Table 9.1.3.

9.1.4 When the dynamic method is used for calculating the seismic action effect of a sluice, the gate chamber shall be considered as a three-dimensional structure.

Table 9.1.3 Dynamic distribution coefficient of seismic inertial force of a sluice

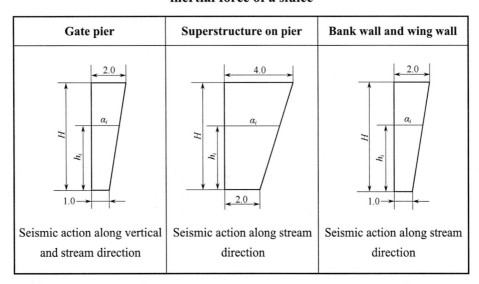

Table 9.1.3 *(continued)*

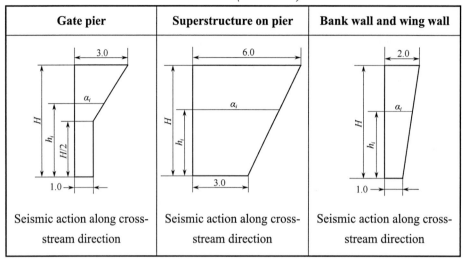

Gate pier	Superstructure on pier	Bank wall and wing wall
Seismic action along cross-stream direction	Seismic action along cross-stream direction	Seismic action along cross-stream direction

NOTES:

1 α_i is taken as 1.0 for the part below sluice bottom slab.

2 H is the height of the structure.

9.1.5 The influence of radial gate rigidity on the seismic performance of a sluice should be calculated, and the dynamic analysis of involved corbels shall be conducted.

9.1.6 In seismic strength check of supports of access bridges and service bridges, the representative value of horizontal seismic inertial force on simply-supported beam supports shall be calculated by Formula (9.1.6) and the representative value of horizontal seismic transverse inertial force shall be borne by supports on both sides.

$$E = 1.5\alpha_h G_{EL}/g \qquad (9.1.6)$$

where

G_{EL} is the characteristic value of structural gravitational action. For fixed supports, the dead weight of superstructure of one bridge span is taken; for movable supports, 1/2 of the dead weight of superstructure of one bridge span is taken.

9.1.7 The representative value of seismic dynamic water pressure acting on a sluice may be calculated in accordance with Articles 7.1.12 and 7.1.14 of this code. When the dynamic method is employed, the dynamic water pressure calculated by Formula (7.1.14) may be converted to an added mass of upstream face corresponding to the unit seismic acceleration.

9.1.8 The representative values of seismic active dynamic earth pressures acting on side piers, bank walls or wing walls of a sluice may be calculated in accordance with Article 5.9.1 of this code.

9.1.9 The structural strength of sluice components shall be checked in accordance with Article 5.7.4 of this code and shall also comply with SL 265, *Design Specification for Sluice*. The influences of structural deformation of sluice component during earthquake on the operation of gates and hoists shall be checked.

9.1.10 When checking the sliding stability along sluice base, the seismic effects shall be determined according to this code and SL 265, *Design Specification for Sluice*. When the dynamic method is used to calculate sluice seismic stability, seismic action effects consistent with that adopted for strength check shall be adopted.

9.1.11 For a sluice seated on bedrock, the sliding stability along its base plane or shallow foundation plane may be checked in accordance with Article 7.1.7 or Article 7.1.15 of this code by the dynamic method or the quasi-static method; When checking the sliding stability of a sluice on soil foundation along its base plane or shallow foundation plane by the quasi-static method, the structural factor shall be taken as 1.2.

9.2 Seismic Measures

9.2.1 When pile foundation is adopted, the connection between pile foundation and sluice base slab and the seepage control measures shall be well designed. Anti-seepage wall, cutoff wall, end sill or other measures may be taken at base slab to prevent piping or concentrated seepage due to the separation of foundation and base slab by seismic action.

9.2.2 The layout of sluice chamber should be well arranged and integrated. The sluice chamber should adopt integrated reinforced concrete structure. Waterstop shall adopt the structural type and material with durability and deformation adaptability. Waterstops at key joints shall be strengthened.

9.2.3 The height of the hoist frame should be decreased through proper type selection and arrangement of gate and hoist to reduce the weight of rack top section.

9.2.4 The hoist frame should adopt frame structure and connection of column of the hoist frame, pier and deck shall be strengthened by increasing section area and reinforcement in the connection positions; When prefabricated concrete beams and non-fixed supports are used for the hoist frame, measures such as guard board, bolt connection or steel clip plate connection shall be adapted for beam support, to prevent fall during earthquake. The stirrup spacing at upper and lower ends within 1/4 clear height of rack column shall be closer.

In the case of design intensity of IX, the stirrup spacing at whole height range shall be closer.

9.2.5 The backfill slope against side pier should be appropriately lowered, and buildings or stacking loads on the bank neighboring to side pier should be avoided, to reduce the seismic deformation of river bank and deformation of sluice due to additional lateral loads during earthquake.

9.2.6 For Grade I, II and III sluices, the upstream blanket should be made of concrete, which should be reinforced properly. The waterstops of contraction joints and drain of sluice downstream and its banks shall be addressed.

10 Underground Hydraulic Structure

10.1 Seismic Calculation

10.1.1 For the underground structures with a design intensity of IX or Grade 1 underground structures with a design intensity of VIII, the seismic safety of the structures and stability of their surrounding rocks shall be checked. For the underground structures with a design intensity of VII and above, the seismic stability of rock mass at portals shall be checked. Seismic safety of Grade 1 underground structures in soil with a design intensity of VII and above shall be checked, and the seismic subsidence of their foundation shall be checked as well.

10.1.2 In the seismic calculation of underground structures, the maximum displacement of the site and its distribution in depth should be obtained from site response analysis. The site is assumed to be a horizontal layered medium and analyzed by the one-dimensional wave propagation method, in which the nonlinear soil model shall be employed. The maximum displacement of bedrock surface may also be calculated through the representative value of design PGA and the predominant period of the site, it shall be halved at the depth of 50 m beneath the bedrock surface or deeper, and may be linear within 50 m.

10.1.3 The seismic effect of underground structures shall be calculated by the displacement response method or acceleration response method. The numerical model shall involve the underground structure and the surrounding medium in a certain range.

10.1.4 For straight tunnel sections in rock, the representative values of axial stress σ_N, axial bending stress σ_M and shear stress σ_V, caused by seismic wave propagation, can be calculated by the following formulae:

$$\sigma_N = \frac{a_h T_g E}{2\pi v_p} \tag{10.1.4-1}$$

$$\sigma_M = \frac{a_h r_0 E}{v_s^2} \tag{10.1.4-2}$$

$$\sigma_V = \frac{a_h T_g G}{2\pi v_s} \tag{10.1.4-3}$$

where

v_p, v_s are the characteristic values of compressive and shear wave velocity of surrounding rock mass, respectively;

E, G are the characteristic values of dynamic elasticity modulus and shear modulus of lining material, respectively;

r_0　　is the characteristic value of equivalent radius of tunnel section.

10.1.5　For straight tunnel sections in rock or soil, the representative values of axial stress σ_N, axial bending stress σ_M and shear stress σ_V, caused by seismic wave propagation, can be calculated by the following formulae:

$$\sigma_N = \max \begin{cases} \beta_N \dfrac{ET_g}{2\pi v_p} a'_h = \beta_N \dfrac{EV_h}{v_p} \\ \beta_N \dfrac{ET_g}{4\pi v_s} a'_h = \beta_N \dfrac{EV_h}{2v_s} \end{cases} \quad (10.1.5\text{-}1)$$

$$\beta_N = \dfrac{1}{1 + \left(\dfrac{EA}{K_u}\right)\left(\dfrac{2\pi}{L}\right)^2} \quad (10.1.5\text{-}2)$$

$$\sigma_M = \beta_M \dfrac{Er_0}{v_s^2} a'_h \quad (10.1.5\text{-}3)$$

$$\sigma_V = \beta_M \dfrac{GT_g}{2\pi v_s} a'_h = \beta_M \dfrac{GV_h}{v_s} \quad (10.1.5\text{-}4)$$

$$\beta_M = \dfrac{1}{1 + \left(\dfrac{EI}{K_v}\right)\left(\dfrac{2\pi}{L}\right)^4} \quad (10.1.5\text{-}5)$$

where

a'_h　　is the maximum value of horizontal acceleration response at the tunnel point of the surrounding mass;

V_h　　is the maximum value of horizontal velocity response at the tunnel point of the surrounding mass;

β_N, β_M　are the reduction factors for axial stress σ_N and axial bending stress σ_M, respectively;

EA, EI　are the characteristic values of axial stiffness and axial bending stiffness of tunnel structure, respectively;

K_u, K_v　are the characteristic values of longitudinal and transverse stiffness coefficient of unit length of surrounding mass, respectively;

L　　is the characteristic value of seismic apparent wavelength.

10.1.6　The seismic effect of underground structures with complicated topographical and geological conditions, such as underground powerhouse, tunnel and other deeply-buried underground caverns, or shallowly-buried caverns such as intake and outlet on river bank, should be analyzed with three-dimensional structural models, in which the dynamic interaction between the structure and surrounding mass is taken into consideration.

10.2 Seismic Measures

10.2.1 Underground structures should be kept away from active faults, shallow and thin ridge, and not be close to hillside or unstable areas. Potentially liquefiable soil should be avoided. The alignment with a greater buried depth is preferred and shall be away from weathered rock mass.

10.2.2 The turning radius of a tunnel and the intersection angle of two tunnels should be large enough.

10.2.3 Undercutting method is preferred when construction conditions permit.

10.2.4 The portals of underground structure should be arranged where topographical and geological conditions are favorable. Measures such as gentle portal slope, shotcrete and bolting or lining protection, and cut-and-cover section should be taken if necessary. The portal structures shall be of reinforced concrete.

10.2.5 The combined effect of lining and surrounding rocks shall be strengthened.

10.2.6 Seismic joints should be set at turning, bifurcation and transition sections with abrupt change in section size or properties of surrounding mass. The number, spacing and structure of seismic joints shall meet the requirements for structural deformation and water sealing.

11 Intake Tower

11.1 Seismic Calculation

11.1.1 The seismic calculation of intake tower shall include the check of stresses or internal forces, overall stability against sliding and overturning, and foundation bearing capacity. Seismic design shall be conducted for nonstructural components, auxiliary electromechanical equipment and their connections with main body of structure.

11.1.2 The seismic action effect calculation of intake tower shall be conducted with the dynamic method or the quasi-static method. The dynamic method shall be adopted to calculate the seismic effect of intake tower with seismic fortification Class A or with design intensity of VIII and above or nonreinforced concrete intake towers higher than 40 m.

11.1.3 The influence of water body inside and outside of intake tower as well as foundation shall be considered for dynamic analysis of seismic action effect of intake tower. The mode decomposition method shall be adopted.

11.1.4 The seismic action effects of intake tower can be calculated by the cantilever method or FEM. But the calculation model shall be the same as that adopted in the analysis of usual load combination.

11.1.5 When the quasi-static method is adopted to calculate seismic action effect of intake tower, the representative value of horizontal inertial force of each mass point shall be calculated in accordance with Article 5.5.9 of this code, where, G_{Ei} is the representative value of gravitational actions concentrated on mass point i of tower body, frame and auxiliary equipment; and dynamic distribution coefficient α_i of inertial forces shall be taken according to Table 11.1.5. When structure height $H = 10$ m to 30 m, $\alpha_m = 3.0$; and when $H > 30$ m, $\alpha_m = 2.0$.

Table 11.1.5 Dynamic distribution coefficient α_i of inertial force of intake tower

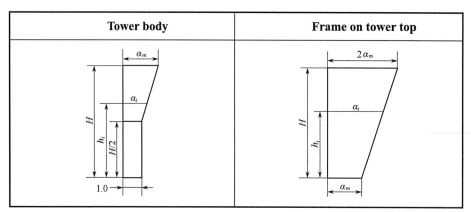

11.1.6 When the dynamic method is adopted to calculate seismic action effect of intake tower, the hydrodynamic pressure inside and outside of tower may be considered respectively as the added mass of the interior and exterior surfaces of the tower, which is calculated by Formula (11.1.6):

$$m_w(h) = \psi_m(h) \rho_w \eta_w A \left(\frac{a}{2H_0} \right)^{-0.2} \quad (11.1.6)$$

where

$m_w(h)$ is the representative value of added mass of hydrodynamic pressure per unit height at water depth h;

$\psi_m(h)$ is the distribution coefficient of added mass, taken as 0.72 for the hydrodynamic pressure inside tower, and taken according to Table 11.1.6-1 for the hydrodynamic pressure outside tower;

η_w is the shape coefficient, taken as 1.0 for the inside of tower and outside of circular tower and taken according to Table 11.1.6-2 for the outside of rectangular tower;

A is the area enclosed by the section-water contact line, where the section is the average section of the tower along height;

a is the mean value of maximum width of structure-water contact surface along the tower height perpendicular to the direction of seismic action.

Table 11.1.6-1 Distribution coefficient ψ_m of added mass

h/H_0	$\psi_m(h)$	h/H_0	$\psi_m(h)$
0.0	0.00	0.6	0.59
0.1	0.33	0.7	0.59
0.2	0.44	0.8	0.60
0.3	0.51	0.9	0.60
0.4	0.54	1.0	0.60
0.5	0.57		

Table 11.1.6-2 Shape coefficient η_w for outside of rectangular tower

a/b	η_w	a/b	η_w
1/5	0.28	1/4	0.34

Table 11.1.6-2 *(continued)*

a/b	η_w	a/b	η_w
1/3	0.43	2	2.14
1/2	0.61	3	3.04
2/3	0.81	4	3.90
1	1.15	5	4.75
3/2	1.66		

NOTE b is the width of tower along the direction of seismic action.

11.1.7 When the quasi-static method is adopted to calculate seismic action effect of intake tower, the representative values of hydrodynamic pressure can be directly calculated by Formula (11.1.7-1):

$$F_T(h) = a_h \xi \rho_w \psi(h) \eta_w A \left(\frac{a}{2H_0} \right)^{-0.2} \qquad (11.1.7\text{-}1)$$

where

$F_T(h)$ is the representative value of the resultant force of hydrodynamic pressure on tower surface per unit height at water depth h;

$\psi(h)$ is the distribution coefficient of hydrodynamic pressure at water depth h, taken as 0.72 for the hydrodynamic pressure inside the tower, and taken according to Table 11.1.7 for hydrodynamic pressure outside the tower.

The representative value of resultant force of hydrodynamic pressure exerting on whole tower surface can be calculated by Formula (11.1.7-2) with its action point at water depth $0.42H_0$.

$$F_T = 0.5 a_h \xi \rho_w \eta_w A H_0 \left(\frac{a}{2H_0} \right)^{-0.2} \qquad (11.1.7\text{-}2)$$

Table 11.1.7 Distribution coefficient $\psi(h)$ of hydrodynamic pressure of intake tower

h/H_0	$\psi(h)$	h/H_0	$\psi(h)$
0.0	0.00	0.3	0.79
0.1	0.68	0.4	0.70
0.2	0.82	0.5	0.60

Table 11.1.7 (continued)

h/H_0	$\psi(h)$	h/H_0	$\psi(h)$
0.6	0.48	0.9	0.20
0.7	0.37	1.0	0.17
0.8	0.28		

11.1.8 When the water depth is different inside and outside the tower, the representative value of hydrodynamic pressure or the added mass at a certain elevation shall be calculated respectively in accordance with inside and outside water depths and then the mean value can be taken.

11.1.9 For tower group connected in a row, when ratio of average width of upstream surface perpendicular to the seismic action direction to the maximum water depth in front of tower a/H_0 is greater than 3.0, the resultant force for the quasi-static method and added mass for the dynamic method, for calculating the hydrodynamic pressure per unit height at water depth h outside of the tower, shall be calculated by the following formulae, respectively:

$$F_T(h) = 1.75 a_b \xi \rho_w a \sqrt{H_0 h} \tag{11.1.9-1}$$

$$m_w(h) = 1.75 \rho_w a \sqrt{H_0 h} \tag{11.1.9-2}$$

11.1.10 Distribution of representative values of hydrodynamic pressure and the added mass on horizontal section can be taken uniformly on both structure-water contact surfaces perpendicular to the seismic action direction for rectangular cross-section tower; and taken as per $\cos\theta_i$ for circular-section tower, where θ_i is the acute angle intersected by the normal direction of point i on structure-water contact surface and the direction of seismic action. The maximum distribution strength of the hydrodynamic pressure and added mass may be calculated by the following formulae respectively.

$$F_\theta(h) = \frac{2}{\pi a} F_T(h) \tag{11.1.10-1}$$

$$m_\theta(h) = \frac{2}{\pi a} m_w(h) \tag{11.1.10-2}$$

where

$F_\theta(h)$, $m_\theta(h)$ are respectively the maximum distribution strength of hydrodynamic pressure and added mass on horizontal section at water depth h. $F_\theta(h)$ acting on both structure-water contact surfaces shall be in the same direction.

11.1.11 When checking the stability against sliding and overturning and foundation bearing capacity of tower under seismic action, seismic action effect by the dynamic method shall be multiplied by the reduction factor.

The seismic sectional bearing capacity of reinforced concrete intake tower shall be checked according to Article 5.7.4 of this code. For seismic check of stability against sliding and overturning and foundation bearing capacity of tower, the seismic action effect shall be in consistent with seismic strength check.

11.1.12 Under the seismic action, the partial factor for rock property of tower foundation may be taken as the static one, but the characteristic value of dynamic bearing capacity may be taken as 1.50 times the static one.

11.1.13 The stability against sliding of intake tower shall be checked by the shear-friction formula.

11.1.14 When checking the foundation bearing capacity of intake tower, the vertical normal stress acting on the foundation surface shall be calculated with the cantilever method.

11.1.15 In seismic check of intake tower, structural factor for stability against sliding shall not be smaller than 2.70. In this case, dynamic shear strength adopts mean value of static one. The structural factor for stability against overturning shall not be smaller than 1.40; the structural factor for foundation bearing capacities of average vertical normal stress and maximum vertical normal stress on foundation plane shall not be smaller than 1.20 and 1.00, respectively.

11.2 Seismic Measures

11.2.1 For the intake tower with high water head and large flow, the earthquake-resistant tube type structure, with high rigidity, strong capacity against overturning, large bearing capacity, and good integrity, shall be selected. For the frame structure, the strength and rigidity of connecting points and bracing members shall be strengthened to ensure the integrity and torsional rigidity.

11.2.2 On the prerequisite of meeting operational requirements, the intake tower structure shall be simple and symmetric, with gentle variation of mass and stiffness, low stress concentration, and sufficient lateral stiffness. Lateral support shall be provided along the tower height, and the support stiffness should be strengthened at abrupt change section.

11.2.3 The tower should be built on the rock foundation with sufficient bearing capacity, buried depth, and consolidation grouting.

11.2.4 For shore-type intake tower, the gaps between the tower and excavated rock mass should be backfilled.

11.2.5 The weight of hoist room on the top of tower shall be minimized. For the seismic vulnerable parts such as pier and the connection between tower and access bridge, such measures as increasing connection area between bridge and the top of tower, flexible connection, and falling prevention of access bridge of hoist during earthquake shall be taken, and the seismic resistance of bridge pier shall be strengthened.

11.2.6 The intake tower group should be lined up and connected to each other so as to increase the lateral rigidity.

11.2.7 Emergency gates shall be provided for Grade 1 and 2 intake towers. The inlet gate slot shall be provided with baffle plates which do not affect ventilation, to prevent debris from falling into the gate slot and affecting the opening and closing of the gate during earthquake.

11.2.8 The seismic measures for the concrete intake tower in detailing, material and reinforcement shall comply with DL/T 5057, *Design Specification for Hydraulic Concrete Structures*.

12 Penstock and Surface Powerhouse

12.1 Penstocks

12.1.1 Seismic action effects of exposed penstock shall be calculated with the quasi-static method. The representative value of horizontal seismic inertial force of each mass point can be calculated by Formula (5.5.9), where, G_{Ei} is the characteristic value of gravity action concentrated on mass point i including water in penstock. Dynamic distribution coefficient α_i shall be taken in accordance with Table 12.1.1.

Table 12.1.1 Dynamic distribution coefficient of seismic inertial force of penstock α_i

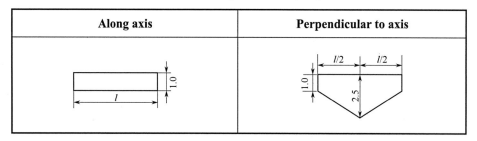

NOTE l is the span between supporting piers of penstock.

12.1.2 The strength and stability of penstock under seismic action shall be checked in accordance with DL 5141, *Specifications for Design of Steel Penstocks of Hydroelectric Stations*.

12.1.3 Seismic check need not be conducted for the penstock embedded in gravity dam.

12.1.4 The penstock shall be arranged on sound rock foundation without abrupt change slope, and kept away from cliff, depression, collapse and landslide. The alignment of penstock should conform to topographic slope direction. The powerhouse shall be prevented from flooding due to adjacent penstock damages during earthquake.

12.1.5 Anchor blocks of exposed penstock shall be placed on bedrock. Foundation treatment shall be done for supporting piers on soil. The span between the supporting piers should be shortened. The cross section and anchor bars should be appropriately increased. The reinforcement should be increased in stress concentration zones.

12.1.6 The ductility of connecting structures of penstock should be increased, to prevent falling down of penstock from supporting piers during earthquake.

12.1.7 Joint and connecting structures at the outlet of penstock embedded in gravity dam shall have a good earthquake resistance.

12.2 Surface Powerhouses

12.2.1 The principle and method for seismic calculation of the substructure of powerhouse are the same as those of concrete gravity dam.

12.2.2 The overall stability against sliding under design seismic action shall be calculated by the shear-friction or friction formula and comply with NB/T 35011, *Design Code for Powerhouse of Hydropower Stations*.

12.2.3 The vertical normal stress acting on the powerhouse foundation surface under design seismic action shall be calculated with the cantilever method. The bearing capacity of bedrock and the tensile strength of foundation surface shall be checked according to NB/T 35011, *Design Code for Powerhouse of Hydropower Stations*. The characteristic value of bedrock dynamic bearing capacity may be taken as 1.50 times the static value.

12.2.4 For the sectional bearing capacity of the superstructure of powerhouse, the seismic check shall be conducted according to Section 5.7 of this code, and the acceleration at the top of substructure shall be taken as the seismic input of powerhouse superstructure.

12.2.5 Joint type and waterstop for the submerged portion of powerhouse shall meet the seismic requirements, and the waterstop material and type should have good seismic resistance.

12.2.6 Seismic measures for the superstructure of powerhouse shall comply with DL/T 5057, *Design Specification for Hydraulic Concrete Structures* and GB 50011, *Code for Seismic Design of Buildings*.

12.2.7 The bank-side powerhouse should be seated on the foundation of stable bank slope and good geological conditions, and the slope against powerhouse should be away from high, steep and dangerous cliff or potentially unstable slopes. The rock slope against powerhouse shall be excavated in stable slope, and shotcrete and rock bolt support should be applied. Protective measures should be provided at the slope side of powerhouse.

13 Aqueduct

13.1 Seismic Calculation

13.1.1 For aqueducts with design intensity VII and above, the seismic action in longitudinal, transversal and vertical direction shall be taken into account simultaneously.

13.1.2 For Grade 1 aqueduct, the dynamic method shall be used for seismic calculation with a three-dimensional model considering effects of adjacent structures and boundary conditions.

For Grade 2 aqueduct, the dynamic method may be used for seismic calculation. The pier and the upper aqueduct body may be modeled as a cantilever and a simply supported beam, respectively.

For Grade 3 and below aqueduct, the quasi-static method may be used for seismic calculation of pier and aqueduct body respectively in accordance with Article 5.5.9 of this code, where the dynamic distribution coefficient α_i of seismic inertial force of pier may refer to Article 9.1.7 of this code, and that of aqueduct body may refer to Article 12.1.1 of this code.

13.1.3 When pile foundation is adopted, effect of pile-soil interaction shall be considered, which may be modeled by equivalent springs of soil, and calculated by the m-parameter method, referring to JGJ 94, *Technical Code for Building Pile Foundations*.

13.1.4 In calculation of Grade 1 or 2 aqueduct, the hydrodynamic pressure within the aqueduct shall be calculated by the formulae in Appendix B of this code.

13.1.5 The dynamic analysis of aqueduct may be conducted by the mode decomposition response spectrum method. For Grade 1 aqueduct, the time history analysis method shall be used according to Article 5.5.8 of this code.

13.1.6 When there is significant difference of geological conditions or abrupt change of topographic features along the aqueduct, the spatial variation effect of seismic input ground motion should be studied.

13.1.7 When the dynamic method is adopted in checking the cross section bearing capacity of pre-stressed reinforced concrete aqueduct body, the reduction factor for seismic action effect should be 1.0.

13.1.8 The calculation of hydrodynamic pressure acting on the aqueduct pier in river channel may comply with GB 50111, *Code for Seismic Design of Railway Engineering*.

13.2 Seismic Measures

13.2.1 For aqueducts with design intensity Ⅷ and above, shock absorption or isolation devices such as lead-core rubber bearing, spherical damping bearing or pot bearing meeting the requirement of bearing capacity should be installed between the aqueduct body and pier.

13.2.2 For the aqueduct with shock absorption and isolation devices, the resonance of aqueduct structure during earthquake shall be considered for the lower supporting structure with low rigidity and on soft soil foundation.

13.2.3 Guard boards shall be installed on top of pier to prevent aqueduct body from lateral falling. Sufficient connection length shall be reserved between pier and aqueduct body end to prevent aqueduct body from longitudinal falling.

13.2.4 The connecting position between aqueduct body end and bearing as well as the top of pile foundation shall be strengthened by reinforcement.

13.2.5 The waterstop type and material meeting the seismic requirements shall be selected for the joint between adjacent aqueduct sections.

14 Shiplift

14.1 Seismic Calculation

14.1.1 Seismic calculation of shiplift tower shall include check of deformation, strength, and overall stability against sliding and overturning.

14.1.2 For design intensity Ⅶ and above, vertical seismic action shall be considered.

14.1.3 Torsion effect under horizontal seismic action shall be studied for structures with nonuniform or unsymmetrical mass or rigidity distribution.

14.1.4 For tower structure not higher than 30 m, the quasi-static method may be adopted for seismic action effect calculation and dynamic distribution factor of seismic inertial force may be taken with reference to Article 11.1.5 of this code.

14.1.5 For tower structure higher than 30 m, the mode decomposition response spectrum method shall be used to calculate seismic action effect. The time history analysis method should be adopted for the calculation of Grade 1 shiplift tower.

14.1.6 For rack and pinion vertical shiplift, dynamic interaction between ship chamber and tower structures and dynamic fluid-solid interaction between ship chamber and water shall be considered. Hydrodynamic pressure on steel ship chamber can be determined referring to Article 13.1.4 of this code.

14.1.7 For dynamic analysis on tower, if the structure is connected with counterweight, the connection shall be modeled by springs with the same stiffness of guide wheel and rail, and dynamic coupling analysis shall be conducted. In simplified analysis, 30 % of counterweight mass may be added to tower column to simulate the interaction between counterweight and tower.

14.1.8 Seismic design shall be conducted for nonstructural components, auxiliary electromechanical equipment and their connecting parts with main structure.

14.2 Seismic Measures

14.2.1 For the shiplift tower, the earthquake-resistant tube type structure with high rigidity, good stability against overturning, high bearing capacity and good overall stability should be selected. Seismic joints should be provided between tower units seated on different types of foundation.

14.2.2 The tower should be regular and symmetrical in shape. The mass, stiffness, and bearing capacity of lateral-force resisting members of same type should be uniformly distributed. The eccentricity between stiffness center and centroid should be minimized. Abrupt change of stiffness in adjacent layers and bearing capacity of lateral-force resisting structures should be avoided.

14.2.3 For the structural design of shiplift, load transfer path under seismic action shall be clear and simple; members and their joints on the transfer path shall not allow brittle failure, and the failure of partial structures or members shall not cause bearing stability failure of the whole structure system.

14.2.4 For rack and pinion vertical shiplifts, damping devices should be provided at the guide mechanism coupling ship chamber and tower.

14.2.5 The non-structural members of floor and roof and non-load-bearing walls at the stairway shall be reliably connected with the main structure to prevent personal injury and important equipment damage by falling during earthquake.

14.2.6 The supports and connecting parts of electromechanical equipment mounted on structures shall comply with relevant seismic requirements.

14.2.7 The seismic measures for the shiplift tower in detailing, material and reinforcement shall comply with DL/T 5057, *Design Specification for Hydraulic Concrete Structures*.

Appendix A Seismic Stability Calculation of Embankment Dams with Quasi-Static Method

A.0.1 When the slip circle method considering inter-slice forces, derived from the simplified Bishop method, is adopted, the representative value of action effect S and the characteristic value of resistance R for seismic stability of dam slope can be determined by Formulae (A.0.1-1) and (A.0.1-2) as shown in Figure A.0.1.

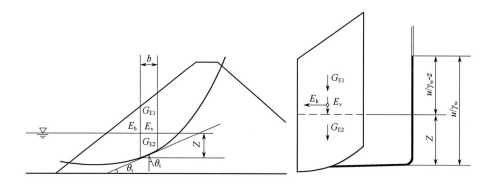

Figure A.0.1 Sketch of slip circle method considering inter-slice force

$$S = \sum (G_{E1} + G_{E2} \pm E_v) \sin\theta_t + M/r \tag{A.0.1-1}$$

$$R = \sum \left\{ \left[(G_{E1} + G_{E2} \pm F_v) \sec\theta_t - (u - \gamma_w z) b \sec\theta_t \right] \frac{\tan\varphi}{\gamma_f} + \frac{c}{\gamma_c} b \sec\theta_t \right\}$$
$$\left[1/(1 + \tan\theta_t \tan\varphi / \gamma_R) \right] \tag{A.0.1-2}$$

$$\gamma_R = \frac{\gamma_0 \psi \gamma_d (1 + \rho_c)}{\dfrac{1}{\gamma_f} + \dfrac{1}{\gamma_c} \rho_c} \tag{A.0.1-3}$$

$$\rho_c = \frac{cb \sec\theta_t}{\left[(G_{E1} + G_{E2} \pm E_v) \sec\theta_t - ub \sec\theta_t \right] \tan\varphi} \tag{A.0.1-4}$$

A.0.2 When the slip circle method without considering inter-slice forces, derived from the Swedish circle method, is adopted, the representative value of action effect S and the characteristic value of resistance R for seismic stability of dam slope may be determined by Formulae (A.0.2-1) and (A.0.2-2).

$$S = \sum \left[(G_{E1} + G_{E2} \pm E_v) \sin\theta_t + M_h / r \right] \tag{A.0.2-1}$$

$$R = \sum \left\{ \left[(G_{E1} + G_{E1} \pm E_v) \cos\theta_t - (u - \gamma_w z) b \sec\theta_t - E_h \sin\theta_t \right] \frac{\tan\varphi}{\gamma_f} + \frac{c}{\gamma_c} b \sec\theta_t \right\}$$
(A.0.2-2)

where

G_{E1}　is the characteristic value of actual weight of the part above corresponding water level in a slice;

G_{E2}　is the characteristic value of buoyant weight of the part below corresponding water level in a slice;

E_h　is the representative value of horizontal seismic inertial force acting on the centroid of a slice, i.e., characteristic value of actual weight of the slice multiplied by $a_h \xi a_i / g$ at the centroid;

a_h　is the representative value of horizontal design PGA;

ξ　is the reduction factor for seismic action effect, taken as 0.25;

α_i　is the dynamic distribution coefficient of inertial force of mass point i;

g　is the gravitational acceleration, taken as 9.81 m/s^2;

E_v　is the representative value of vertical seismic inertial force acting on centroid of a slice, i.e., characteristic value of actual weight of the slice multiplied by $a_h \xi a_i / 3g$ at the centroid, in the acting direction of upward (−) or downward (+) whichever is unfavorable to stability;

M_h　is the moment of E_h to the circle center;

r　is the radius of the circle;

θ_t　is the included angle between slip circle radius through midpoint of slice bottom and vertical line; for the slice at the dam axis side of the vertical line through the center of the circle, plus sign is taken; on the contrary, minus sign is taken;

b　is the slice width;

u　is the representative value of pore water pressure at midpoint of slice bottom;

z　is the vertical distance from corresponding water level to midpoint of slice bottom;

γ_w　is the unit weight of water;

c, φ　are the internal cohesion and friction angle of soil under seismic action, respectively;

γ_0 is the structural importance factor, taken according to GB 50199, *Unified Design Standard for Reliability of Hydraulic Engineering Structures*;

ψ is the design situation factor, taken as 0.85 according to Article 5.7.1 of this code;

γ_E is the partial factor for seismic effect, taking 1.0 according to Article 5.7.1 of this code;

γ_c, γ_f are the partial factors for material properties of soil shear strength, respectively. $\gamma_c = 1.2$, $\gamma_f = 1.05$; partial factor for material properties of non-linear friction angle of coarse-grained materials like rockfill and gravels can be taken as $\gamma_f = 1.1$;

γ_d is the structural factor.

Appendix B   Calculation of Hydrodynamic Pressure in Aqueduct

B.0.1   In seismic calculation of Grade 1 aqueduct, the hydrodynamic pressure acting on aqueduct body with rectangular or U-shaped cross section may be divided into impulsive pressure and convective pressure, as shown in Figure B.0.1.

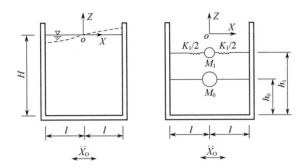

(a) Aqueduct with rectangular cross section

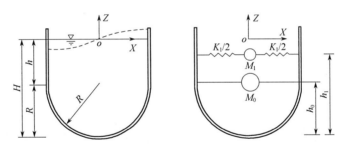

(b) Aqueduct with U-shaped cross section

Figure B.0.1   Sketch of hydrodynamic pressure model

1   Impulsive hydrodynamic pressure

Under transverse seismic action on aqueduct, the impulsive pressures distributed in aqueduct may be converted into horizontally added masses fixed on each side wall along elevation direction and calculated using Formula (B.0.1-1) for $|z/l| \leq 1.5$ and Formula (B.0.1-2) for $|z/l| > 1.5$, respectively.

Translator's annotation: The "H/l" in the Chinese version of this code has been corrected to "|z/l|" in this English version, and Formula (B.0.1-1) corrected accordingly.

$$m_{\mathrm{wh}}(z) = \frac{M}{2l}\left[\left|\frac{z}{H}\right| - \frac{1}{2}\left(\frac{z}{H}\right)^2\right]\sqrt{3}\tanh\left(\sqrt{3}\frac{l}{H}\right) \qquad (\text{B.0.1-1})$$

$$m_{wh} = \frac{M}{2H} \tag{B.0.1-2}$$

For aqueduct bottom, when $H/l \leq 1.5$, the impulsive pressure can be calculated by Formula (B.0.1-3).

$$p_{bh}(x,t) = \frac{M}{2l} a_{wh}(t) \frac{\sqrt{3}}{2} \frac{\sinh\left(\sqrt{3}\frac{x}{H}\right)}{\cosh\left(\sqrt{3}\frac{l}{H}\right)} \tag{B.0.1-3}$$

where

- M is the total water mass per unit length along aqueduct axis direction, taken as $2\rho_w Hl$ for rectangular cross-section and $\rho_w(2hR+0.5\pi R^2)$ for U-shaped cross section;
- $a_{wh}(t)$ is the horizontal acceleration response at center of aqueduct bottom of each cross section;
- ρ_w is the density of water;
- H is the water depth in aqueduct;
- $2l$ or $2R$ is the internal width of aqueduct.

When $H/l > 1.5$, the impulsive pressure at aqueduct bottom is distributed linearly.

2   Convective hydrodynamic pressure

Under transverse seismic action on aqueduct, the convective pressure may be considered as spring-mass system connected with aqueduct side wall at height h_1. For aqueducts with rectangular cross section, the equivalent mass M_1, equivalent spring stiffness K_1 and height h_1 can be calculated by Formulae (B.0.1-4) to (B.0.1-6), respectively.

$$M_1 = 2\rho_w Hl \left[\frac{1}{3}\sqrt{\frac{5}{2}}\frac{l}{H} \tanh\left(\sqrt{\frac{5}{2}}\frac{H}{l}\right) \right] \tag{B.0.1-4}$$

$$K_1 = M_1 \frac{g}{l}\sqrt{\frac{5}{2}} \tanh\left(\sqrt{\frac{5}{2}}\frac{H}{l}\right) \tag{B.0.1-5}$$

$$h_1 = H \left(1 - \frac{\cosh\left(\sqrt{\frac{5}{2}}\frac{H}{l}\right) - 2}{\sqrt{\frac{5}{2}}\frac{H}{L} \sinh\left(\sqrt{\frac{5}{2}}\frac{H}{L}\right)} \right) \tag{B.0.1-6}$$

For aqueduct with U-shaped cross section, the equivalent mass M_1, equivalent spring stiffness K_1 and height h_1 can be calculated by Formulae (B.0.1-7) to (B.0.1-10), respectively.

$$M_1 = M\left(0.571 - \frac{1.276}{\left(1+\frac{h}{R}\right)^{0.627}}\left[\tanh\left(0.331\frac{h}{R}\right)\right]^{0.932}\right) \quad \text{(B.0.1-7)}$$

$$K_1 = M_1\omega_1^2 \quad \text{(B.0.1-8)}$$

$$\frac{R}{g}\omega_1^2 = 1.323 + 0.228\left[\tanh\left(1.505\frac{h}{lR}\right)\right]^{0.768} - 0.105\left[\tanh\left(1.505\frac{h}{R}\right)\right]^{4.659} \quad \text{(B.0.1-9)}$$

$$h_1 = H\left\{1 - \left(\frac{h}{R}\right)^{0.664} \times \left[\frac{0.394 + 0.097\sinh\left(1.534\frac{h}{R}\right)}{\cosh\left(1.534\frac{h}{R}\right)}\right]\right\} \quad \text{(B.0.1-10)}$$

Under vertical seismic action, only impulsive pressure may be taken into consideration. For aqueduct bottom, the impulsive pressure may be taken as uniformly-distributed added mass fixed on it and calculated by Formula (B.0.1-11):

$$m_{wv} = 0.4\frac{M}{l} \quad \text{(B.0.1-11)}$$

For aqueduct side wall, the impulsive pressure may be considered as horizontal pressure distributed along elevation and calculated by Formula (B.0.1-12). The impulsive pressures on both side walls of aqueduct at each instant are in the same direction.

$$p_{wv}(z,t) = 0.4\frac{M}{l}a_{wv}(t)\cos\left(\frac{\pi}{2}\frac{H+z}{H}\right) \quad \text{(B.0.1-12)}$$

where

$a_{wv}(t)$ is the vertical acceleration response at center of aqueduct bottom of each cross section.

B.0.2 For Grade 2 aqueduct, in calculation of transverse seismic action effect of aqueduct pier, the mass of aqueduct body and added mass of hydrodynamic pressure of two adjacent half spans shall be taken as additional concentrated mass at pier top.

In calculation of aqueduct structure, when $H/l \leq 1.5$, the impulsive pressure under horizontal seismic action may be taken as additional concentrated mass in transverse direction at depth h_0 on aqueduct side wall and calculated by Formulae (B.0.2-1) and (B.0.2-2), respectively.

$$M_0 = M \frac{\tanh\sqrt{3}\frac{l}{H}}{\sqrt{3}\frac{l}{H}} \qquad (B.0.2\text{-}1)$$

$$h_0 = \frac{3}{8}H\left\{1+\frac{4}{3}\left[\frac{\sqrt{3}\frac{l}{H}}{\tanh\left(\sqrt{3}\frac{l}{H}\right)}-1\right]\right\} \qquad (B.0.2\text{-}2)$$

When $H/l > 1.5$, horizontal added mass uniformly distributed on aqueduct inside wall below the point of $|z| = 1.5l$ can be calculated by Formula (B.0.1-2), and linearly distributed impulsive pressure acting on aqueduct bottom may also be corrected accordingly.

Effect of convective pressure may be considered as spring-mass system connected to aqueduct inside wall at height h_1. For aqueduct with rectangular cross section, the equivalent mass M_1, equivalent spring stiffness K_1 and height h_1 can be calculated by Formulae (B.0.1-4) to (B.0.1-6), respectively. For aqueduct with U-shaped cross section, they can be calculated by Formulae (B.0.1-7) to (B.0.1-10), respectively.

The acceleration response at pier top shall be taken as the seismic input for bearings at the bottom of aqueduct body.

Explanation of Wording in This Code

1 Words used for different degrees of strictness are explained as follows in order to mark the differences in executing the requirements in this code.

    1) Words denoting a very strict or mandatory requirement:

       "Must" is used for affirmation; "must not" for negation.

    2) Words denoting a strict requirement under normal conditions:

       "Shall" is used for affirmation; "shall not" for negation.

    3) Words denoting a permission of a slight choice or an indication of the most suitable choice when conditions permit:

       "Should" is used for affirmation; "should not" for negation.

    4) "May" is used to express the option available, sometimes with the conditional permit.

2 "Shall meet the requirements of…" or "shall comply with…" is used in this code to indicate that it is necessary to comply with the requirements stipulated in other relative standards and codes.

List of Quoted Standards

GB 50011,	*Code for Seismic Design of Buildings*
GB 50111,	*Code for Seismic Design of Railway Engineering*
GB 50199,	*Unified Standard for Reliability Design of Hydraulic Engineering Structures*
GB 50287,	*Code for Hydropower Engineering Geological Investigation*
GB 18306,	*Seismic Ground Motion Parameter Zonation Map of China*
NB/T 35011,	*Design Code for Powerhouse of Hydropower Stations*
DL/T 5016,	*Design Specification for Concrete Face Rockfill Dams*
DL/T 5057,	*Design Specification for Hydraulic Concrete Structures*
DL 5141,	*Specifications for Design of Steel Penstocks of Hydroelectric Stations*
DL/T 5353,	*Design Specification for Slope of Hydropower and Water Conservancy Project*
DL/T 5395,	*Design Specification for Rolled Earth-Rock Fill Dams*
SL 265,	*Design Specification for Sluice*
JGJ 94,	*Technical Code for Building Pile Foundations*